Dr. Adventure

DANGER AND DISCOVERY
FROM POLE TO POLE

✳ ✳ ✳

By

Warren M. Zapol, M.D.

ISBN: 979-8-218-78812-4 (paperback)
ISBN: 979-8-218-78811-7 (hardcover)

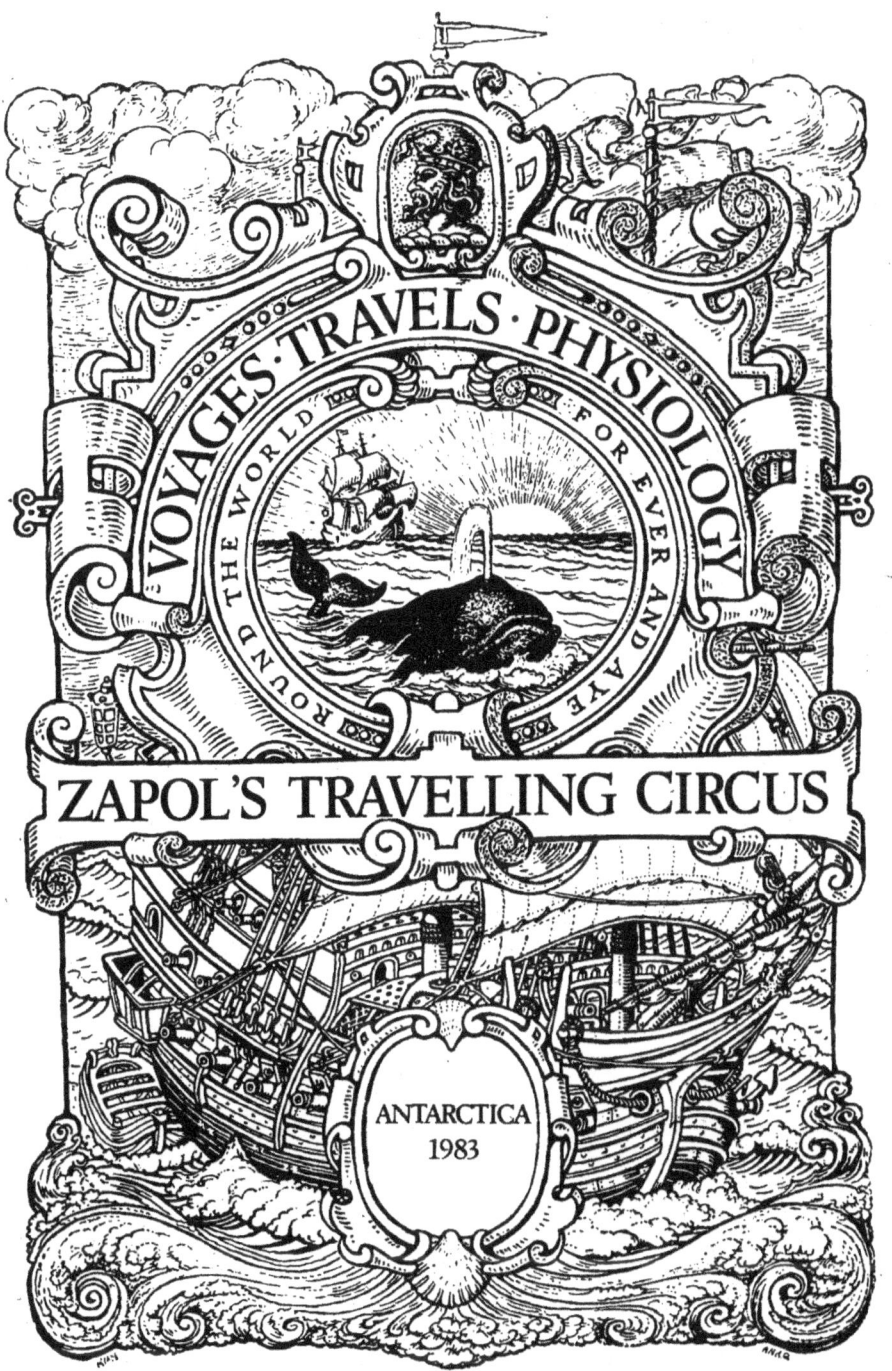

From the Zapol Antarctic Expedition, 1983.

Table of Contents

* * *

CHAPTER 1

Baby's Breath

✳ ✳ ✳

IT IS NOT A GOOD sign when babies turn blue after birth. It's right and proper to intervene and help a baby in this situation. It's vital. No matter what else is happening, blue is a sign that a newborn baby's lungs aren't working the way they should, that they aren't receiving the right signals at the right time in the right way. Babies need to go from being divers—swimming in amniotic fluid inside their mothers, letting mama's lungs do the work—to being climbers, making strides on their own, opening up arteries and sending blood to their lungs to breathe on their own. This is a process I've spent my whole life studying, from inside the intensive care unit in Boston to below the ice of Antarctica. This baby was born blue, flown in on an emergency helicopter to lie on this surgical table, and her body was failing.

A small baby weighs about as much as a chicken. Imagine a chicken on an enormous surgical table, surrounded by a team of nurses and surgeons. They are prepping the skin with alcohol, getting ready to cut. This team was trained for just that, for cutting, and that afternoon in 1990 you could feel the inevitability of the next actions in the room. Press blade to skin, make incision, insert tubes. Pray.

I had come to question the inevitable, and that day to join Jay Roberts, a tall, calm anesthetist and neonatologist who was working with me on a new treatment for blue babies, and so I stayed in the corner of the room, out of the way. I never trained as a Neonatal Intensive Care Unit (NICU) doctor, I am just an inventor and anesthesiologist. When it came time to translate my ideas into practice, we needed a whole team to make it work. Jay was charming even in his blue and white scrubs, even after days and nights of working in the NICU, and his good nature was useful in this moment. Jay walked over to the huddle of surgeons. "Excuse me," he said. "We have a permit to do an experiment first."

We showed them the cylinder of gas, which fortunately didn't reveal the skull and crossbones on the side. We'd papered over it, so that her mother and father wouldn't think that we were killing their baby girl with poison gas. I thought it also might help with the surgeons that we'd disguised the skull and crossbones.

It didn't. The surgeons turned and glared at us over their masks.

"Ugh, you're wasting our time. Let us just go ahead and put 'em on ECMO."

The baby had been flown to our hospital, not for our experiment, but to receive extracorporeal membrane oxygenation (ECMO).

I stepped over from the corner. "Look," I said. "ECMO might work. *Might.* We think *this* will work better. Please, let us try it, it won't take more than half an hour."

I helped develop ECMO at the National Institutes of Health (NIH) 20 years earlier. The machine was designed to pull blood out of the baby's vein, diffuse oxygen into the blood, and then return oxygenated blood to an artery in her neck. Our hope had been that it would be the end of the story—the baby would be fine. Except we found from the very beginning of our testing ECMO in humans that the procedure had a high likelihood of causing brain bleeds, not because of any fault of the surgeons, but simply because of the blood thinners required.

I could feel the exasperation. And I knew the surgeon's view of the world— I had started out in a surgical residency, and I knew from their point of view anesthesiologists are just supposed to make sure the patient is asleep and alive while the surgeons let their knives fly. The baby had been flown in for ECMO, and they wanted to give ECMO.

Until that moment in 1990, ECMO was the best we could do for babies whose lungs simply refused to turn on when they were born. But I knew we could do better. Brain-bleeds leading to long-term disability could result in a terrible life for this newborn.

Throughout my career, I have tried to save the lives of the mortally wounded, the sick-unto-death. I led the Anesthesia Center for Critical Care

Research at the Massachusetts General Hospital (MGH) and spent 50 years in the laboratory studying intensive care. The critical moment near death is a key moment for invention, for creativity, because there are no good options, and we can only help and learn and do better. Each time I think, "What if we try this?" It is better than thinking nothing at all. Then we experiment, which is the best we can do at that moment. That is our job, to buy them time so they can get better. For many of our patients, it is just too late. They are going to die. But some of them do recover.

It was the death of a young girl in 1988 that kicked me into action and sent me off searching for a better solution than ECMO to add oxygen to the blood while reducing blood pressure in the lungs. She must have been fourteen, with pulmonary hypertension, and I was treating her with prostacyclin, a newly discovered drug at the time. Clinicians thought that prostacyclin might help patients like her. So we tried it, because she had no other options. I gave it to her intravenously. Her pulmonary pressure did come down, as we had hoped. It seemed like success! But only for a moment… then her blood pressure dropped, too. Her heart stopped, and we could never start it up again.

Even if a person is that sick, on the brink of death, it is a weighty moment when an experiment hastens their demise. We are intervening in a natural process, trying to stay a shadow that is falling across a patient. But in trying to slow it, sometimes we hasten it. It's a risk we take when we practice medicine especially when we're searching for new and better treatments. I deeply believe we have an obligation in these tragic moments to learn what we can, to gather information that will lead to better approaches in the future. I hope that my death will do the same for science. I recognize that death is the greatest cost, not just to patients, but to their families, and it impacts the people like me who care for them. Before going on we always stop and reflect, write and discuss, and learn from our experience so that we can do better by the next patients we serve.

So, after losing the girl, I was driven to find a new way, and went back to the books.

The gas that we were considering for the baby was nitric oxide (NO). Nitric oxide was discovered in 1772 by Joseph Priestly and is as simple a molecule as you can imagine, with just one nitrogen atom and one oxygen atom. For centuries after Priestley's discovery, NO was classified as a poison.

Perhaps you've had an experience with nit*rous* oxide, which we call "laughing gas." Priestly also discovered that molecule, which has not one but two nitrogen atoms, and again one oxygen atom. And about 25 years later, a daring English chemist, Sir Humphry Davy, began experimenting with inhaling the gas, leading to laughter. In 1799, he famously shared it with the poet Samuel Taylor Coleridge, who wrote:

> The first time I inspired the nitrous oxide, I felt a highly pleasurable sensation of warmth over my whole frame, resembling that which I remember once to have experienced after returning from a walk in the snow into a warm room. The only motion which I felt inclined to make, was that of laughing at those who were looking at me.

In 1844, Horace Wells would begin to use it for dentistry, which is how many of us know it today.

1990, 156 years later. Now the surgeons gave us the okay to use the nitric oxide on the baby on the surgical table.

The baby? Who was this baby? Why was she flown in by helicopter? Did she have siblings? What jobs did her parents have? What was her room like?

I didn't know, and it wasn't my concern.

Another doctor might describe the human drama, the *Grey's Anatomy* plotline that led to the hospital. But that's not the medicine I practice. For me, the first and in some ways the only job was to understand what is going on with a patient's body. The way to do that best was to set the drama of life aside and look only at the facts in front of me. That worked for me in Antarctica when someone fell into a crevasse and the rescue team was en route. That worked for me when I was hand-pumping oxygen into a dying cruise passenger while flying over the Transantarctic Mountains. And that worked for me when I was in the hospital with a critically ill patient. I saw the body: the tubes and pipes and valves and pumps that needed to work for life to continue. I cared above all about the mechanisms of oxygen transfer that kept the baby breathing. I wanted it all to work. I wanted to hear this baby give a full-lunged "Waaaaah." I wanted to see this baby turn pink. But this baby had blue skin, and wasn't making a sound.

So we opened the valve on that nitric oxide welding gas tank. It had been a century and a half since a new gas had been discovered for medicine. That's how rare this kind of opportunity is.

I remember many people saying, "Doctor Zapol has lost it." Skepticism was everywhere, and it wasn't limited to that operating room. Even one of the scientists who discovered the natural production and use of nitric oxide within the body had told me, "Warren, you can't inhale nitric oxide, it's a poison!" Paracelsus, the great German-Swiss physician and philosopher, is paraphrased as saying, "The dose makes the poison." Water can save your life or kill you, depending on how much you drink. Compounds from the leaves of the yew tree are used to fight cancer, yet those same leaves people eat to commit suicide.

There were lots of reasons for reasonable people to think I was wrong. There is a disease called silo filler's disease which is an overdose of nitric oxide that occasionally is observed in farmers who are entering silos filled with gas from fermentation. Another case of accidental overdose was featured in a gripping 1967 article in the *British Journal of Anaesthesia*. Two patients undergoing gynecologic procedures were anesthetized with a batch of nitrous oxide that had been contaminated with NO. They got cyanosis, meaning, they turned blue ("cyanosis" comes from the Greek *kyanos*, meaning dark blue). Even their blood turned blue. Here's what their anesthesiologist said:

…slight cyanosis was observed… cyanosis deepened and was not improved by ventilating with 100 per cent oxygen. The patient then started to look extremely ill…. the next patient on the list was sent for… She immediately started to become cyanosed and it was then, at once, obvious that this was a case of poisoning, but poisoning by what we had no idea…. the pathology laboratory reported that the abnormal pigment was definitely methaemoglobin. Because of this report both patients were given an intravenous injection of 10 ml of 1 per cent methylene blue and in each case the colour turned to bright pink within about 90 seconds. After this the second patient improved considerably and her blood pressure rose to normal.

That first patient was not so lucky. She died.

Nitric oxide binds to hemoglobin and forms what the pathologists saw in her blood: methemoglobin. Methemoglobin is hemoglobin that is unable to carry oxygen, and that is why her blood turned blue, ultimately leading to her death. The blue blood that we observe in babies that are born without functioning lungs is also caused by a lack of oxygen. Throughout the 20th century and until today, nitric oxide poisoning was known to be reversed by treatment with the drug methylene blue. Methylene blue helps the hemoglobin recover more quickly from that poisoning. As the first synthesized medication, it was used for treating a number of other diseases including malaria. But soldiers in the tropics complained of green or blue urine, and the whites of their eyes turned strange shades of blue. They refused the treatment; this is what we call a lack of "acceptability" of a medicine. As a result of these disconcerting side effects and the refusal of patients to accept them, methylene blue is now rarely used beyond the infrequent methemoglobin treatment.

That is not the only risk with NO. NO is converted, or oxidizes, into nitrogen dioxide, NO_2, a pollutant, when it reacts with oxygen in air. When nitrogen dioxide combines with water, it forms nitric acid. You've heard of acid rain. Well, we risked creating acid rain in the lungs. So it is no surprise that each cylinder of NO that we bought was made for welding, because it certainly wasn't made for hospitals—and it was stamped with the skull and crossbones we had covered up, together with a label detailing the toxic properties of inhaling NO:

NOT FOR HUMAN USE: Can cause death or permanent injury after a very short exposure to small quantities. Irritant of eyes, nose, throat; can cause unconsciousness. Nitric oxide forms acids in the respiratory system which are irritating and cause congestion in the lungs.

I always have the image of the skull and crossbones in my head when working with NO, a palpable reminder that I am holding the balance of life and death in my hands.

For me, invention is the result of piecing basic units together, like nitrogen and oxygen. Electricity. Fire. Things we live with every day. I've been very focused my whole life on tinkering with these basic components of the universe. When you mess with these elemental forces, the unexpected happens. Like in the Captain Marvel stories I devoured as a kid—where Billy Batson was transformed into a superhero by lightning. "Shazam!" I too was inspired to play with elements, to see what I would become.

When I was in middle school in the 1950s, I had a steady supply of comic books, provided by my friend Leon whose father had a subway newsstand in 14th Street Union Square. We'd huddle around a little pot-bellied stove in the L train station to keep warm while we chose our comics. We could take the comic books on a Friday if we'd return them on Monday. It was like a library! He'd pack 'em up again and sell them off on Monday morning. So I read a lot of Fawcett Comics. They published Captain Marvel, which was my favorite because my dad's business was manufacturing and selling Captain Marvel watches.

There were four of us: myself, Leon Greenfield, Stanley Green, and Marty Rich. These were the days when the space race was just heating up, and the four of us were determined not to be on the outside looking in. Aside from our Marvel super-heroes, we devoured the science fiction of Willey Ley and Isaac Asimov. We were sure that we, too, had what it took to be rocket-men. We dreamt about space travel, and we thought about going to the moon, and we started to build rockets. We wanted to be just like Werner Von Braun. Except we were Jewish.

On TV there was a crazy scientist, John Stapp, who had a show where they conducted scientific experiments live, like the *Mythbusters* of the 1950s. In one episode he tested the effects of rocket acceleration on the human body. There were shots of Stapp's face as he was on a rocket on a railroad car. He would accelerate along, and his whole face would get splattered back. He would routinely break his bones. He was dubbed "The Fastest Man Alive" when he hit 632 mph on tracks on December 10, 1954. We thought, "Wow! We want to do that!"

Stapp used railroad cars, so we decided we would take Stanley's little Lionel railroad set and set it up in his backyard. Behind all these row houses, you might have a 500-foot stretch of backyard, which is where we laid down the track. You see, I had been paying close attention in class, to make my mother proud. And with my electronics knowledge, and Marty's chemistry know-how, we had all we needed to make a rocket. We got a 50-millimeter shell, drilled it out, and filled it with zinc sulfide. We knew we needed a venturi to accelerate the burning gas. The brass ring of a doorbell has just the right angles on it, so we found an old doorbell, and bolted the brass ring with screws and nuts to the shell. We had a dilemma then. How would we ignite it? So we got a nickel chrome wire, and attached it to a Lionel transformer, and thought we'd put the power through it to ignite it. So we set it up in the backyard along maybe a 20-foot stretch.

We decided we'd stop it with a cardboard box. We got down into the blockhouse, which was the cellar. I threw the switch, nothing happened. So I figured, well, the nichrome wire wasn't insulated, and was probably shorting to the doorbell venturi, it just needed to be lifted up.

I went out to the rocket, lifted it up, and that's the last thing I remember.

Next thing I knew, I was lying on the ground. A lady was hanging out a window and yelling in Italian, something about a rocket and her building. It didn't make sense until it did. The rocket flew off our track and kept flying and slammed into her wall. The cardboard box didn't stop it.

I couldn't see very well—my eyes were teary. Then I couldn't see at all. I was almost blind. People were looking at me, but they didn't see me. They saw a little gray man, with little gray eyebrows and tufts of gray hair. I had burnt off my eyebrows and my hair on the front of my head.

My friends guided me home. I lived around the corner, and they pressed the doorbell and ran away. When my grandmother opened the door, she took a minute to realize that the little gray man on her doorstep was me. She gasped, and swiftly escorted me to the New Lots bus which took an interminable half hour to arrive at Dr. George Meister's office. I was plenty frightened, shaking, and afraid of permanent blindness. Dr. Meister was my favorite family doctor, and also a member of the family, a cousin of my mother's. He checked my eyes and looked into my cornea.

"What did you use as rocket fuel?"

"Zinc Sulfide." I said meekly. My grandmother tsk-tsked.

Dr. Meister laughed. The little gray man smiled, though he didn't know why.

"Zinc Sulfide?" Dr. Meister said. "That's an eye ointment! Nothing to worry about, you'll be good in a couple of days. Go home with your grandmother."

Through my dazed state, I had a thought: the power of chemistry was just astounding. You could make rockets or you could make ointment with zinc and sulfur. Or, like me, you could make both!

So I went home with my grandmother. Somebody said, the worst words in the world to hear as a child are: "When your father comes home…" But the threat was always worse than the eventuality, at least with my dad, who loved me unconditionally.

That wasn't the only time I tangled with explosives. I also played with gas in long tubes in the lab at NIH when we wanted to treat Vietnam war soldiers whose chests and lungs had been destroyed by bombs. When I sit down to invent, I put everyone else's work aside, and I reason from first principles with what I know to be true from the books, and from the times things have (literally) blown up in my face. And somehow, all of what I know and had worked on for decades, since I was a child, since I was at NIH, came together when it came time for me to find a better way to treat lung failure.

Inspiration is a funny thing. People might imagine it to be the preserve of artists, writers and the ilk, but it's quite key for scientists, too. In 1988, I visited Los Angeles to look at taking a leadership role at the University of California, Los Angeles (UCLA), and had the good fortune to spend some time with Lou Ignarro, a sprightly, mustachioed Italian-American chemist. Lou's warm personality and brilliance in *his* research on NO focused *my* research squarely on the potential therapeutic applications of NO for the respiratory disorders I had devoted years of my life to researching and battling. He and I discussed the fact that NO was very much a deadly compound, and he was the first of many to warn me against giving it directly to patients, instead advocating for other methods of delivery. Indeed I was grappling with how to prevent the toxicity of inhaled NO in a therapeutic context, which could be as deadly as the British patient experienced with the contaminated cylinder. It was a good conversation, and rewarding to be in the presence of someone who not only asked the right questions but had some answers. Lou became a lifelong friend and in 1998 won the Nobel Prize for Physiology or Medicine.

I was sitting in the LAX airport lounge on my way home, leafing through the newspaper when I noticed that the *Los Angeles Times* weather forecast included statistics on the airborne levels of CO, NO_2, SO_2 and various particulates. There they were again, nitrogen, oxygen, sulfur, and the all-pervasive carbon which was well on its way to building up in the atmosphere. The problem of emissions was such a huge issue in LA and the surrounding valleys, that for decades it had been considered an essential public service to inform readers of air pollution levels. I reasoned that NO must have been produced by internal combustion in cars and buses and then oxidized in air to NO_2. But this oxidation is not an instantaneous process. So it made sense that people had to be breathing NO in cities, particularly in tunnels, and perhaps even more at rush hour, where I had just been, sitting in LA traffic.

Paracelsus' adage was running through my head throughout the flight, and immediately upon my return to MGH I was making the leap to separate the beneficial from the toxic effects of nitric oxide. Building on Lou's work, we had a good idea that nitric oxide is a natural gas produced by the body to signal the blood vessels to open or close. I called Washington and found the people at the Occupational Safety and Health Administration (OSHA) who regulate what we breathe at work.

I said, "I'm a doctor at Mass General Hospital, and I'm interested in how much nitric oxide I can breathe safely."

The nice woman replied, "Well, you can breathe 25 parts per million of nitric oxide for eight hours each day while you fry hamburgers at McDonalds, or any other workplace."

My heart leaped. Aha! I could breathe NO safely! I recalled that this was about the same amount that the newspapers were reporting in the Los Angeles basin. Twenty-five parts per million means the air contains 0.0025% NO! In this field of work we use parts per million (ppm) rather than percentages. So we were talking about *tiny* quantities of NO in the air here, but what I learned is that we could safely breathe *some*.

I thanked her and hung up.

With this number in mind, I could imagine safely conducting experiments in the lab using a similar level of the gas. If you could breathe it, I reasoned, it might dilate the blood vessels of the lungs, but it would instantly be destroyed by the hemoglobin in your red cells and never circulate around your body, because the hemoglobin would convert it to nitrate or nitrite, the same stuff that they preserve bacon with. You'd pee it out. But the lungs would work. The dose also makes the medicine.

The office I was calling from was in an old building that felt more like a charming Parisian attic than a modern hospital office, overlooking the roundabout at MGH where ambulances would come in at all times of the day, bringing critically ill patients to our ICU. My wife, Nikki and children Liza and David would come to pick me up from work sometimes late in the evening, and I could wait until they were outside, and then run down at the last minute, so I could squeeze in the review of another scientific paper, or revise a grant or chat with a colleague. Throughout this period, my research was supported by funding from the National Heart, Lung, and Blood Institute (NHLBI) of NIH, and guided by an incomparable cohort of clinicians at Mass General, and I needed to work very hard to keep that support.

On that particular day I was distracted from the frenzy of activity outside, however, cutting up a brown paper bag and wrapping up the cover of a copy of *The Satanic Verses*, a Salman Rushdie novel that had caused quite a controversy, including a fatwā calling for the assassination of the author. At this time, David was sixteen and developed sciatica in his back from playing viola in the local youth orchestra. He ended up in the hospital for a steroid injection into his back and a week in traction to relieve the pressure on the disc. I was wrapping this book up so that he could read the latest from his favorite author without raising any security concerns, when a Swedish doctor, Claes Frostell, stuck his head into my office.

A blonde Swede, Claes had a cherubic look that was charming and fit with his kindly bedside manner. An anesthesiologist from Uppsala, he had recently come to Boston to work in my lab. I said, "Claes, come with me, let's go visit David." On the walk down the irregular old stairs that looked like an Escher drawing, Claes and I discussed his interests. Claes wanted to study pulmonary hypertension, the focus of our lab, and went through a few ideas for new areas of research. I listened as best I could, dodging and weaving through the doctors in the hallways, rolling beds on their way to surgery, and waving hello to colleagues as we found our way to David's room. I delivered the book, and visited for a while.

Seeing David back in the hospital brought up memories. It's one thing to treat other people's kids, but David was born six weeks prematurely, like Albert Einstein, and I'll never forget watching his breathing in the incubator as we hoped and waited for him to come home from the hospital. It was Nikki's uncle who went at that time to the market, bought a chicken that was about the size of David, 3lbs 13oz, and marveled at the fact that he might possibly live, even at that size.

On the way back to the office, I was still thinking about babies and turned to Claes, and suggested the best model we could find for babies. "Get a sheep," I said. We work a lot with sheep in our lab. Sheep don't bite and they don't bark and we're pretty good at measuring their pulmonary circulation. I lowered my voice to a whisper. "I want to try to give it NO to breathe." Claes' eyes widened. "We'll give it something to tighten up its pulmonary circulation and see if we can reverse it."

I'm a teacher, I've taught hundreds of students in my laboratory and in my courses, lectured to tens of thousands of physicians and scientists in my career. I often refer to them as my intellectual children. You never know how you are going to influence people. Something in that conversation lit up Claes, and within a week we came up with a test and sought approval from the hospital's animal research oversight committee for the protocol. It was soon granted.

But it wasn't an easy procedure, or one without risks. Our fears were twofold: There was a chance, at least theoretically, that we would rapidly convert the circulating hemoglobin to methemoglobin like what happened to the poor women in the 1967 story in the British Journal of Anesthesiology. That would kill the sheep. Secondly, we were concerned by the possibility that NO would cause lung injury, by transforming to nitric acid, which could also be lethal to the lamb.

We were optimistic, because you have to be optimistic, as well as curious, in any experimental setting, but we couldn't have anticipated in our wildest dreams just how well this would go—from the initial moments of the procedure, we could see it would be a knock-your-socks-off first experiment, totally unforgettable.

Someone lowered the music on the radio in the lab, a Mozart symphony, because this was a moment for focus. Madonna was popular at that time, which my ten-year old daughter Liza insisted on blasting at home, but the lab was more respectful. They knew when I was coming and switched to 99.5, Classical Radio Boston. I had a subscription to the Boston Symphony Orchestra since college, and I was never a fan of pop music after the Beatles had their moment.

Claes and I slid the gas masks over our faces, and Jay opened the windows of the lab—something that had clearly never been done before, as it took a hammer and screwdriver to get it to open. We watched the sheep calmly breathing through a tracheostomy to deliver the gas right into the airway. We gave just a few parts per million of nitric oxide and as we opened up the tank, we watched the numbers on the measurement device go up: 5, 10, 20, 30, 40, 80.

And guess what? Bingo! Within minutes pulmonary pressure went down, but the systemic blood pressure didn't change. This is the key to it—inhaled NO only worked in the lungs of that sheep, not in the bloodstream. After turning off the NO gas for a few minutes, we found that the pulmonary pressure increased. We were thrilled by this reversibility—I remember looking into Jay and then Claes' eyes through the gas masks and seeing them as wide as my own. We knew we were close to success, very close. But was it safe?

Yes: Ten minutes after breathing NO, the methemoglobin levels were still low. The inhaled NO had produced the desired effect of dilating the blood vessels in the lung without reducing blood pressure in the body, and had not produced significant levels of methemoglobin. With repeated trials, eventually we would learn that several hours of breathing NO was safe and did not produce lung injury or nitric acid in our lambs.

This was beyond exciting. It was truly exhilarating. There could be many possible clinical uses for inhaled NO.

But we had a serious hurdle ahead. Before this new therapy could see its prime time it would need to be tested in humans. We had to make a choice, as three applications immediately came to mind: newly born infants with persistent pulmonary hypertension, children with congenital heart disease and pulmonary hypertension, and adults with acute respiratory distress syndrome (ARDS). All were seriously ill patient groups for whom few medical options existed and for whom it made physiological sense to try this new pulmonary vasodilator strategy. Unfortunately, most of these patients were critically ill with high mortality rates, which meant the human trials to show the efficacy of NO would be very risky. We didn't want the trials to cast doubt on the effects of NO; these patients were suffering from the failure of multiple organs, not just the lung. We chose babies.

Why did I think it might work in babies? Well, the clock that makes you turn on nitric oxide production is inside you, and all babies have it. If we look at mouse pups, we see that the enzyme to make nitric oxide doesn't appear until two, three days before the mouse is born. So there's a clock in the baby that says, "Oh, you're going to be born in two or three days, you better make this nitric oxide, cause you're going to need to turn on circulation to the lungs." You don't need it when you are swimming in utero, it's all fluid, you're as good as a fish, you're not breathing. But on Day One you take a deep breath and you better breathe, and there better be blood flow in that lung to pick up oxygen and bring it to your brain and muscle and make you pink. Otherwise you'll be blue. And blue is a good color for the sky, but not for blood or babies.

That's why the lung's nitric oxide machine has to kick in. When we treated a lamb, we supplemented the lamb's natural NO-generating machine with inhaled nitric oxide from a can. And it worked. We hoped that the natural enzyme that makes NO in the body would turn on after a few days. At which point we hoped the lamb would survive just fine, and live the rest of its life pink. So the idea of inhalation of NO is simple: we're supplementing a natural substance that wasn't there in certain newborns.

But what we knew so far was that it worked in a lamb. We had yet to show that we could do anything for a human.

If you are going to try something in people, generally you have to ask the Food and Drug Administration (FDA), so I started by calling and asking if they would allow us to begin a human trial at MGH in near-death babies with pulmonary hypertension. The FDA representative who spoke to me on the phone reasoned that NO was a gas like an anesthetic—after all, wasn't I an anesthesiologist? "All the gasses are grandfathered in," she said, "you can go right ahead." I still wonder to this day whether she confused NO with nitrous oxide, N_2O, laughing gas. Was this a regulatory error in my favor?

Next we had to get approval from my hospital. Because I had tested it already in the lab with Claes and Jay, we were able to convince the Human Studies Committee at our hospital to allow us to try it in humans within a year of our first tests in sheep. The Human Studies Committee kind of knew me; they knew me as a serial, responsible, inventor, that I'd do something in the lab, and then if it was really good I'd want to try it in people. And so they'd see me coming and say, "Ok, he doesn't kill people, he's very careful, they're making clean experiments, they don't hurt people." It has served me through the years to be methodical and careful and build the case stepwise. It happened again during the COVID-19 pandemic when we treated pregnant mothers who had COVID-19 with inhaled NO, which was important because they were not eligible for other therapies.

We were overjoyed when our proposal was approved for treating babies. We had been worried about having to go through more animal studies, possibly having to try it in adults first, and anticipated appeals, consultations, and so much red tape. Of course, we weren't given a carte blanche either; the initial approval granted by MGH permitted only a half hour of NO exposure in the babies. We weren't going to look a gift-horse in the mouth, though. This was our chance to start what was a world first: clinical trials of inhaled NO in pediatric patients.

I remember sharing my excitement with Nikki. She had long become a true believer in my efforts, and she anticipated our success. I have been very lucky to have such constant love and support. I could never have done any of this without her.

Even if you have spent your whole life studying medicine and physiology and have tested your ideas on animals and even lambs and filed patents and believe in your heart of hearts that you are doing the right thing for a baby, it is still a leap of faith to expose a baby to an unknown. I'm a scientist, my faith is in the data, and it took a year of data to get to the point where I was willing, along with Jay and Claes and all our colleagues, to give nitric oxide to a baby. To that fragile newborn child on the table, whose parents had trusted us with her care, to repair her body and rescue her from lung failure and death.

So there we were, in the corner of the room with an enormous machine, and we had all the permits and permissions to try this big thing on the tiny baby girl. We had done all the tests on sheep. We had cleared this with the hospital. Her parents had signed off, informed of the risks and potential benefits. I thought to myself, "It's human research, so you have to be brave. You have to understand the human disease, you have to believe it will make a dent in the human condition, you have to balance risk and benefit. Can I hurt people? You bet. Just like that fourteen-year-old. Unfortunately." And by saving the most gravely ill first, I had a chance to work my way to the many other people who have less severe illnesses like chronic lung disease and hearts that aren't quite keeping pace to get them up the stairs. At the time, our human research was a leap of faith, but now? Well, now you would be sued if you didn't wheel out the cylinder with the skull and crossbones to save the baby.

I will admit that our apparatus did look cumbersome for a tiny baby, like something conjured up in a tech class rather than something sleek designed in one of the best hospitals in the world. The size of two refrigerators, whizzing and whirring and dials sticking out, with a Mr. Coffee box on top to make it look less intimidating. Some might have looked at it and thought it was a mad scientist's invention. But in truth, there was nothing mad about it. We were trying something truly revolutionary, yes, something that really made people, smart people, scared, but that wasn't the machine. It was because we were marching right into the unknown.

And as luck would have it, this baby was just right, not so sick that she was going to die if treated, but sick enough that she might die if she was not treated. Nitric oxide is not holy water, it's not going to reverse inevitable processes, we need patients that can still be saved. She was the first patient where we said, 'Let's see if she turns pink. Let's not measure anything except how pink she is, just with a little pink meter, a pulse oximeter.' Like the one you put on your finger to test for COVID-19, except it was hooked up to a screen because we wanted to make sure everyone in the room could see if the baby's lungs were still working. Same thing.

That little girl began to breathe a little poison gas, a dose of a natural substance identical to the one made by her own body. Her chest moving fast, like a butterfly, we could start to see the changes before our eyes.

We proceeded, carefully adding stepwise increments of NO to her breath. The response came stunningly quickly and bowled me over. The pulse oximeter showed the blood oxygen level rising from about 50% to 90% within a minute. The blood oxygen levels kept rising as we gave the baby increasing NO doses, mirroring the effect we had seen in lambs. With each turn of the dial, we could literally see the hue of the child's skin change from blue to white; by the time we got to 80 ppm, the maximum we had agreed with the hospital, we had a pink baby on our hands.

The results for the baby were beyond our wildest hopes.

To our amazement, to our relief, to our joy, the baby's methemoglobin levels, just like the lamb's, remained low after this short treatment, meaning that the baby's blood could sustainably carry oxygen.

The chest X-ray confirmed the child had clear lungs and no lung disease. We had selectively dilated her pulmonary circulation, allowing more blood to circulate through her lungs and carry oxygen to the body and its organs. The most remarkable effect was that the dilation appeared to be lung-selective, as we had hoped. The baby's systemic blood pressure did not decrease while breathing NO. Her heart didn't stop. It kept on beating and for all I know it still beats today. That was a resounding success.

I remember the nurse coming up to us afterwards and asking, 'What did you do to that baby?'

I said, "We just gave her a teeny dose of nitric oxide, a poison gas, and it seems to have worked."

And the baby was still pink. So we left the tank on for half an hour, and when we turned it off, the baby stayed pink.

And the surgeons said, "Damn it." Packed up the scalpels and walked out of the operating room.

I'm a critical care doctor. Most of my patients are so sick we never get a chance to talk with them, and so we don't build relationships like the cancer doctors in the hospital, or the cardiologists, or the psychiatrists. Most of my patients are not conscious. But in the case of NO, I was contacted by the parents of many patients over the years. I would go to conferences in places like New Orleans or Tokyo or Paris and parents, many of them doctors, would walk up to me and say "You saved my child, I just want to thank you." And I would smile and ask them about their story and hear how the children grew up over the years and became part of the world.

These babies, hundreds of thousands of them who were saved, made me think a lot about how important that bond is between mother and child. As I would go out to raise support for our inventions, for making nitric oxide available around the world, for trials in China or Uganda, I came to notice that the people who came to support our work were people who had had sick babies themselves. It made me think a lot about our family—about what it was like for Nikki and me to have a sick baby ourselves. It made me think a lot about my own mother, and my childhood in Brooklyn and what made me into a physician-scientist-family man. An adventurer. An inventor.

CHAPTER 2
My First Breath

* * *

As a child, I collected stamps. I would inspect every letter that came through our house and extract the valuable stamps carefully and put them in my albums. I have a particular memory from when I was eight, late at night, under covers with flashlights in my room. Steven and I were looking over my stamps, and he noticed I had a whole page of Hitler stamps—there were all these awful Hitler stamps that came from envelopes that carried letters from when my father spent time at Medical School in Bonn, Germany during the 1930s, before he was expelled as a foreigner and a Jew.

Steven, my first cousin, was a whole year older than me. I didn't have any siblings, so he was like an older brother to me. He was much more of an athlete—he had so much grace—and my mother adored him! I, on the other hand, couldn't hit the baseball to save my life, and was always the last picked when my classmates made a team.

Steven stared at Hitler's ugly *punim*, and, imitating our grandfather Harry, spat on the ground. Harry, who was sleeping soundly upstairs, while he and Grandma Gussie babysat for us.

Not to be outdone, I spat too, but my aim was off and I hit my foot, which I then rubbed on the rug.

Clearly, this wasn't a sufficient illustration of our hatred for Hitler. What could we do?

We decided we would burn Hitler in effigy. In miniature. Steven carefully crafted this little raft of toilet paper and popsicle sticks. I then peeled a Hitler stamp out of its little vellum case, and placed it on the raft, which we then set to sea—in the toilet. Since we thought we might have trouble getting it to light, we poured some of my dad's Zippo lighter fluid on it.

We dropped a match on our little Hitler pyre. "Boom!" It went up in flames, filling the room with smoke. My mother, who loved pink, had plastic pink flowered curtains hanging over the window. They burst into flames—and all the smoke rose up through the center shaft. You see, our bathroom was connected to grandma and grandpa's bathroom upstairs, through a little center air shaft up to a skylight. My grandparents smelled the smoke.

"Warren? Steven!" I heard them yelling down the shaft.

"Fire! Fire!" We yelled back, panicking, throwing towels on the melting curtains.

My grandfather came rushing down, and threw a pot of water at the fire, and another, until the fire hissed down and my mother's plastic curtains were a toxic pink puddle.

My grandparents never got angry at us, and they protected us from my parents. "It was an accident!" They explained.

Maybe it was the Hitler thing—who could blame us for wanting to burn the guy? Or maybe, probably, it was because they loved us so much, even though we were naughty. There was this sense in my family that I could do no wrong. Of course, in Jewish families, sons are treated like princelings. But for me it was different, and I didn't understand why until later. Steven and I were very, very embarrassed, of course. Sometimes you feel guiltier when you are not punished. I would have to make up for it by working extra hard at school.

Our home was 393 Atkins Avenue in East New York, Brooklyn. Ours was a two-story house on a block of two-story houses, owned by my grandparents, upstairs. I lived downstairs with my parents. My mother's parents were Jewish immigrants from Polish Galicia, then part of the Austro-Hungarian empire, and had fled a generation before the Second World War and persecution of the Jews in the Third Reich. Growing up, I heard the sounds of Yiddish filtering down the stairs from their lively conversations. The sounds of Yiddish mingled with the smells from my grandmother's kitchen, where she cooked every meal. The kitchen itself was a rich mystery; there was always something on the go. Dumplings, stuffed cabbage, pickles, chicken legs hanging out of the sides of soup pots—I have to confess I was relieved that these weird appendages were reserved for my father.

With my father Ben in Brooklyn circa 1944.

My first memories are going to school. My mother, Florence, was a petite, blue-eyed, blonde substitute school teacher at PS 202 just a few blocks from our home. She would walk me down to the school, her thin hand holding mine, and we would both head into the building. She was so very careful about the teachers that I had, and about my performance. When I brought home a test that was graded 99, she would ask, "What happened to the other point?"

I seem to remember being fearful as a child, maybe this was because my mother was a bit fearsome with her high standards, and also because she was so watchful for danger and missteps, like I was made of some kind of precious material. I started out too skinny and easy to get picked on, and as I got older, I got plump and was an even easier target—the short, curly-haired, dark-eyed goody-goody who sat in the first row, first seat. I had a particularly difficult time controlling my bowels when I got anxious, and I remember being taken home several times, at arm's length to be washed up and cleaned up. Those were my earliest memories, from kindergarten and first grade.

Playing violin accompanied by my mother Florence, 1951.

My mother was a pianist, and she directed school plays at PS 202. She would always give me a role. Whether or not I deserved it, I had lines to memorize. She was also president of the PTA, and very close to the principal of the public school, and she guided my education for those first six years very carefully. She was born in 1909, on the Lower East Side of New York City. She had an elder brother Jack, and a younger, Sol, and mom was the middle child. She went to high school and teacher training school, but her brothers never went to college.

My father Bernard, or Ben, as he was called, was less involved in my daily life. He traveled a lot to Switzerland, importing Swiss movements for watches. Then he started the Marvel Watch Company, and he made Captain Marvel and Mary Marvel watches, and for a short moment became very successful. When I was eight, in 1950, my father was making a million dollars a year. We had a Cadillac, and we became a very wealthy middle-class family. And yet, within a few decades, most of that money was gone. Despite all his attention on business, my father didn't make the wisest investments. My mother was dismayed with him and looked to me to make better choices with my career, to give us all more security.

My father marketed and sold watches like these featured in a 1948 advertisement.

My mother wanted me to learn to play an instrument, and she picked violin. Everyone Jewish wanted their son to be Jascha Heifetz, the Russian teenage prodigy whose American debut was at Carnegie Hall. She accompanied me to violin lessons, and she always practiced with me. I can still hear her thumping on the key when I got the note wrong, oh boy, she had a great ear. My teacher, who lived in the area, swelled with unabashed pride when I was accepted to study with Raphael Bronstein, the famous violin teacher in Manhattan, a chain-smoking Russian immigrant with an ashtray absolutely full of cigarette butts. Even better that he had studied in the same violin class as Heifetz in Russia! He made me play Vivaldi concertos, and I practiced consistently, every day. Yes, it felt like a chore, but I was driven to persist. By my mother. And over time, this drive became a part of me. I needed to practice because it scratched an itch and occupied my very active mind. Even now, I practice my violin every day. I need to.

In my spare time, I turned to radio, where I soon developed a lasting fascination. My parents encouraged this habit since it was less destructive than burning stamps or shooting off rockets, even after I fitted an antenna to our Brooklyn rooftop and my transmitter destroyed TV reception for the whole block. At that time, televisions were extravagances that only the richest could afford. There were just a few of those on our block for my radio to screw up. *Whiskey Norway 2, London Queen Ocean…Whiskey Norway 2, London Queen Ocean.* They would hear a voice buzzing on their televisions and scrambling the image. That was me, calling out with my new FCC call sign WN2LQO.

I loved joining the worldwide fraternity of amateur radio operators which included some of my best friends from school. There was no internet obviously, and long-distance telephone calls were out of the question; at ten dollars a minute to connect to another continent, my corner of the world felt very confined, and the larger world unreachable. Yet, as a result of World War II, when thousands of radio operators had been trained across the world for military purposes, there was then a proliferation of stations just like mine around the world. And, what's more, because of the sunspot cycle which was peaking in the 1950s and 60s, amateurs like me could bounce radio waves off the sky with precision, allowing our signals to cross the continents, traverse oceans and timelines. All of a sudden, the world opened up to me; I had friends everywhere. We became pen pals! Each time I made radio contact, we would exchange special postcards, called QSL cards. I have a collection of thousands of these colorful postcards, from Tanzania to Tasmania, India to Antarctica.

I became adept at communicating in Morse code, much in the way that kids these days can compose text messages in a flash. But voice communication was also possible; for example, I had a regular lunchtime chat with a lobster fleet captain out of Cape Town, South Africa. My curious young mind feasted on these images of adventurous lives in far off places, and very luckily I was indulged and encouraged by my family. I felt I could travel the world, and in a sense, I did just that. I could hear the English broadcast of Soviet propaganda on Radio Moscow; and then with a spin of a dial, I could listen to the BBC or Voice of America. In the space of a decade, I made contact with about 200 countries in the world, and each time, I would locate and label the country on my world map on the wall. Radio makes an entire world reachable.

In 1955 and at the age of thirteen, I went to Stuyvesant High School, and it continued to broaden my horizons: it was a school that super smart nerd boys like me had to test into, and I loved it. I brought Marty Rich with me to the test, and he also got in, so it was great to have a friend sharing the experience. When we went it was still on the wartime sessions, meaning they tried to get twice as many kids through the school as usual, and they pushed me through in three years. My first year, class started at noon and went until five. Then, in the second and third years, I started at seven am and went until noon.

I got into trouble at Stuyvesant, I had my first experiences with being bad at something. It was French, I was awful at it; I was flunking. My French accent through my Brooklynese was unintelligible, and I didn't have the time and patience to learn a foreign language. Marty and I had the same teacher, Mr. Abromovitz, who was a humorless little guy. He would call out in French, in a very nasal voice, "Tout d'abord au tableau noir, d'avant la classe, Monsieur…" There was a very long silence… "Zapol!" Snickers and sighs of relief from the rest of the class as I walked to the blackboard, where some English phrase was written on the board which I was supposed to translate but instead stared at like a dummy.

But my mother salvaged me, she said, "Get a tutor, you need help. If you don't know something and you can't pick it up by reading it, get someone to help teach you." She got help. I think that was also contributing to the death of my violin lessons, because if I was going to understand French, I couldn't be practicing Vivaldi for an hour or two every night. So, academics won.

My time at Stuyvesant High School in Manhattan made a huge impression on me. I was making an hour and a half commute each way from Brooklyn, so I was no stranger to travel, or at least that's how it felt at the time. Every morning was an odyssey; just remembering where to change trains and which exit to take was a challenge.

While my earlier experiments with rockets were earthbound and horizontal, all of a sudden, a vertical rocket exploded onto all our minds, introducing space travel. When I was in my third year at Stuyvesant, in October 1957, the Soviet Union launched Sputnik, the first man-made satellite into earth's orbit. Now, I realize that to the reader today, this may seem just like some quaint little nugget from history. And it's true that there wasn't a whole lot of technology in Sputnik, in and of itself. But its impact would inspire—and in some ways—define a generation. I can still hear the beeps from its radio emissions. It was clear that the Cold War wasn't going to be like the war that came before, but that it would also be ideological and a competition of potential: played out through technological inventions in defense and discovery with missiles and rockets. And the Soviets had the jump on us.

As you might imagine, with my passion for radio, inventing, and my earlier interest in rocketry, I was particularly enthralled by the idea of a radio transmitter going around the world over our heads, and what might happen next. It fueled my interest and passion for science, and I was encouraged and indulged in all quarters, at home and in school. This was very much a childhood in the shadow of Sputnik.

My father, Ben Zapol, came from a family of Russian Jews. They left Europe in 1922 during a period of persecution that predated the Nazi Third Reich, and having left Russia they didn't leave with a good impression of the place. Combine that with the atmosphere of intense anti-Soviet sentiment that was pervasive at the time as manifested in McCarthyism. People were legitimately afraid of being caught in the ideological war on home territory. Like today, we were in uncharted territory, and nobody knew what was coming next.

Suddenly, growing up in the shadow of Sputnik didn't seem like such a good thing. The launch itself was greeted with celebration: Look what humanity is capable of! I suppose that especially for the generations that had lived through the war, this kind of peacetime achievement was a beautiful thing. But the second guessing was soon to follow. I'll never forget my physics teacher telling us to "learn physics or learn Russian." Sent the shivers right up my back, I'll tell you, even if I didn't fully grasp the magnitude of the Cold War, or even what the Soviet Union was all about.

Just the prospect of failing at another foreign language was enough to send me running to physics class. Besides, Stuyvesant High School didn't even offer Russian!

At the end of Stuyvesant, I was 16, and I got into the Massachusetts Institute of Technology (MIT) and I got my driver's license. I earned my freedom just at the right time, because all of a sudden, my family home started to become suffocating. My mother was very clear that Cambridge, where MIT is, was too far from Brooklyn, and she wanted me to go to college closer to home, in New York. I could see her whole plan, and I couldn't tolerate being controlled by her, or anyone.

I remember Marty and I sneaking off in my father's Cadillac for a joy ride in the Catskills. I wanted to get away, as far as possible, the first of a lifetime of escapist adventures. As I vented, Marty confessed to me that my mother never liked him, that she seemed to discourage our friendship. What right did she have to choose my friends, my school, my every step? As we zoomed over the rolling hills, I knew I would choose to go to MIT regardless. I could feel my future ahead of us, my past in the rear view mirror.

It wasn't until 1978, that I discovered that what I thought was a relatively "normal" childhood wasn't at all. That the memories that I understood weren't exactly as I remembered. That everyone was holding on to a big secret, and the only one who didn't know was me.

I was thirty-six, working at MGH, when the call came.

Shirley, my secretary, said to me, "Warren, there's someone who says he's your brother on the phone."

"Bullshit," I said. "I don't have any brothers. Hang up." And she did.

The man called back. "This is not a joke," he said. "I'm his brother, tell him to talk to me."

I finally spoke with the man.

"Who is this?" I said.

"Thank God I finally found you!" he said.

"Who am I?" I said. Not "Who are you?" I wanted to know who he thought I was.

"You're my younger brother," he said. "You're thirteen years younger than I am. My father is Nat Warshaw. My name is Stanley Warshaw, and you're supposed to be Michael Warshaw, but your mother died when you were born."

"This is baloney," I said. I would have said "bullshit" but the man was a stranger. "No one ever told me this story, that I was adopted, that I wasn't a Zapol."

He persisted with credible facts: "Our mother, Millie Warshaw, who was a Russian immigrant, was quite young when she had you. She had eclampsia, high blood pressure, and had a stroke, four days after you were born. Your biological father, Nat, didn't want another baby to raise. He sent me to Los Angeles to relatives to get rid of me, because it was a time of great angst in his life. You were given away to the family across the street who wanted a baby, the Zapols, because Nat was a friend of that family."

"A friend how?" I asked.

The man continued. "Well, he lived on Atkins Avenue too, as I said. And Nat was friends with Sol Rothlein—Florence Zapol's brother."

My uncle. My street.

"Can I call you back later?" Enough for now. I needed to take a breath. I needed to think.

"Sure. Just call me at my wallpaper school." A wallpaper school? What was a wallpaper school? What was happening?

I took his number and hung up.

I was at work so I couldn't show much emotion, but my entire world was reeling. The foundation of my entire existence was called into question. Who am I and who are you? Who the hell was he to call me and change everything all at once? Why now?

I called Nikki in Concord.

She answered, surprised. I would never call in the middle of the work day. In fact, I almost never called from work, period. I dealt with home problems when I was at home, if I had to, and work problems at work. When they mixed, I got insomnia.

"Warren? What is it?" I could hear that she'd been sleeping, with our two-month old baby girl at her side. We had returned home from England a month before, at the end of my sabbatical leave from MGH and Harvard. A year before, David, then five years old, Nikki and I left for Auckland, New Zealand. Nikki and David stayed there, the closest they could get to me while I led a six-week Antarctic expedition. The baby was an "ice baby:" conceived on a weekend I spent on leave from the ice, and born in Oxford, England.

"I got a very strange call…" I filled her in on the details of the call, and Nikki quickly woke up. It was all so surreal. Nikki is a smart lawyer who had a CIA spy for a father, so I hoped she could debunk the strange phone call, and make this whole disruption disappear.

She was full of questions about the call. "What did this man sound like? How come he is so much older than you? What does he know about you, about us? So your name would have been Michael Warshaw?" I'm sure she wasn't pleased about the prospect of a stranger upending my entire identity, and thus, her entire conception of our family, our children, her in-laws, me.

Nikki was likely both familiar with and unnerved by this experience. She had spent much of her childhood in the Philippines, secure in her belief that her father, Gabriel Kaplan, worked for a Non-Governmental Organization (NGO). She only learned about her father's true profession when she was well into her teens, when Gabe's case officer sat down with Nikki and told her the truth about her upbringing, her childhood, shaped by the nascent CIA. Just like me, Gabe had seemed stable and transparent, and now everything was becoming shrouded.

"Nikki, let me give you his phone number and see what you make of this."

To say that a request like this from me was unprecedented would be an understatement. I rarely felt so out of my league. She agreed to call and we hung up.

And I waited.

I picked up a copy of *The New England Journal of Medicine* and read the same sentence over and over without registering a word. I was heeling over, and I hoped, desperately, that Nikki could right the ship.

Back at work, my phone rang. I answered right away, before Shirley could pick up.

"So….?" I asked.

"I think it is real," Nikki said, soberly.

"How? Why? What happened when you called?"

"I was really hoping it was a scam. When I called the number you gave, they answered saying 'Hello, U.S. School?' Which sounded like a very strange cover. Like "the agency." But when I asked 'U.S. School of what?' They replied "U.S. School of Paperhanging," which lined up with what Stanley told you."

"Nick, get on with it. What happened next?" I could sometimes be impatient with long-winded stories. This was one of those times. My life hung in the balance. At least it felt that way.

Nikki took a deep breath. "I was eventually connected with Stanley Warshaw, and introduced myself as your wife, and quizzed him every way I could think of. About his intentions in contacting you, his age, his parents, your parents, your neighborhood. His story was the same he told you, and it correlates with the facts of your past—at least what we thought of as facts. Warren, it's totally incredible, but not impossible. I think it's real. How could he know all those things if what he says isn't true?"

I felt sick to my stomach. "I'm coming home right now," I said, hoarsely.

I left work early and drove home to Concord. I never left work early. I can think of one other time that happened, when David had his head smashed by a golf club during third grade recess, and the local hospital left dirt in his wound before stitching it up. Another mess I helped clean up. I wanted to clean up this mess, but I was starting to feel like I was becoming the mess.

I walked into our bedroom, the late afternoon light streaming through the window, and sat on the bed. I held my head in my hands. "Who am I? How can we know if it really is true?"

Nikki stood before me, holding our baby. "There is one person who can tell you for sure."

Of course. The most logical person. What was I thinking?

I called my mother, in South Beach, Miami, Florida, where she lived then. My father had died four years earlier.

"Warren! What is it, my dear?"

"Mom? I got the craziest call, from a man who said he was my brother."

Silence.

"Mom? I got a call from a man who said he was my brother."

A longer silence.

"Mom. Am I your son? You can tell me if it is true."

I waited. Please, please say yes.

"Ask the man they say is your father," she said tersely.

"What?"

"Ask the man they say is your father," she repeated louder. Like she was angry with me for asking at all.

I was bewildered. I could hear her answer, and I heard it the first time, but I couldn't understand the way she was answering. Why did I have to ask a man, a stranger, about my place in the world? I needed her to be straight with me. Now, more than ever, this really mattered.

Pained, I looked at Nikki. I glanced at her slight figure, and her tiny swollen postpartum belly, holding our baby. "Mom. Did I come out of your body?"

Another silence.

Finally, "No. No, you didn't."

And then I heard her cry.

So it was true. I was adopted. To say that this experience was head-scrambling is to put it lightly. I stared at the wall, then I sunk my head into my hands.

As a doctor, my entire world revolves around the body. A body whose experience of the world is largely informed by its biology, composition, and genetics. I had spent my adult life studying the biological markers for disease and sickness, becoming an expert in the care and maintenance of the body. I learned that if you listen closely enough to the patient, and to their medical history, then there is a truth to the physical body. And if you are good enough, you can help this body.

So now I learned that my understanding of how *my* body came into the world, and how *my* biology was informed by my genetics, was completely false. My mother's deep discomfort and shifting of responsibility to my birth father, a man I never knew, made me so confused. It seemed like she was trying to hurt me, to push me away before I could reject her. Was she angry with me for asking her, after all she had done for me, sacrificed for me? It was as if she felt like I had intentionally disrupted her belief (and illusion) that she was my birth mother.

But the truth is not an opinion, it is not held by a mysterious man who says he is my father. It is a physical fact. I felt like a baby, born from no womb, floating in space. Who did I belong to?

I flashed back to my own associations with adoption. My parents were very careful whenever a friend of theirs had an adopted child. They would be very gentle and kind to that child, they would be very special and they encouraged me to do the same. But they never intimated that I might not be me, Warren Zapol. It amazed me. How had they kept up that pretense so well? Later, when I was twelve or thirteen, my mother wanted to take in a foster child. We had interviews, and I was interviewed by the foster care agency. But there weren't enough foster children at the time, in our part of Brooklyn, so we never had one. My mother had a penchant for children, and she liked to bring them up, and she liked to have them sing and dance and do the things that she did. Her desire for a foster child made sense to me. And even amongst the interviews for a foster child, they never said a word to me about my own adoption. Why hadn't she told me before? Was the truth so horrible that it had to be obscured for so long, perhaps forever, if Stanley hadn't contacted me?

Now that I held this crucial piece of information, that I had indeed been born of different parentage, who was I now? Was I still Warren Zapol? What made me Warren Zapol, and not Michael Warshaw, and what set the two of us apart? What happened to Michael Warshaw as a baby, and where did he go? Did he die in childbirth, or was he, in fact, me?

I looked at myself in the mirror. At this point, I was already a husband and a father of two children, all who bore the Zapol family name. They all knew me as Warren Zapol. Warren Zapol who grew up in Brooklyn, was raised in a two-family house with his parents and grandparents, went to PS 202 and Stuyvesant and then to MIT.

But "my brother" said that I was actually Michael. Was that true? To me? Who might I have been if I had stayed Michael?

*   *   *

The next weekend he came to visit us. Three of them came on Saturday morning: Stanley, his wife Leah, and his younger daughter, Charlene, about six. Stanley was a large man, tall, with a swagger. His skin was olive, like mine. He had big, bushy eyebrows, like mine. His frame was like mine, but I was thirteen years younger, less rough-hewn. It was uncanny.

Stanley strode into our house as if he belonged there and plunked down at our kitchen table, waiting for breakfast. He never asked a single question of me, or Nikki, except possibly cream for his coffee. He told us of the long saga of all the difficulties he had gone through to find me. When he had a heart attack a year ago, he decided to find me before he died. He wrote to his Vermont State Senator to ask for help, figuring my last name was Zapol. The senator searched the Public Health Service for Zapol, and followed a long trail which led to calling me at Mass General.

"Goddamned search of a lifetime!" He told me. "I can't believe I did it. My own fucking brother! I found you before I croaked." He slapped me on the back, a little too hard.

I didn't know that I was lost, and I hadn't asked to be found.

It was a few days before Halloween. After breakfast, Charlene and David, two kids oblivious to the significance of this event, went out on the deck to carve a pumpkin.

Leah took a photo out of her pocketbook and handed it to Stanley. Stanley held the image out to us. "Here is a photo of your mother, Warren." It was a black and white photo of a man and a woman on a beach, maybe Coney Island. I stared at the photo. She had my jaw line, gentle smile. I recognized my feet. She was beautiful.

"Is that your father next to her?" Nikki asked.

"Yes."

I was confused. Why hadn't he said: This is a picture of your parents. Or our parents? Was it because she was dead and he was alive?

He went on: "That would have been shortly after they were married."

There was no question that the man in the picture was my father. He had my slumped-in shoulders, my hairy chest.

Looking at the picture, I suddenly did feel lost. I needed something.

"Can I keep this?" I asked.

Leah said, "Of course. It's for you."

Stanley looked away. I had the feeling he didn't want me to take the picture. What did my birth family owe me? Apparently nothing.

Stanley told us more. He was thirteen when his mother was pregnant. His father was not around much at the time. Though Stanley knew his mother had not been feeling well, he was shocked to never see her again after she went to the hospital to deliver the baby. She disappeared. Just like the baby.

He heard that his baby brother was given away to neighbors across the street. Soon after, his father sent him to live with an aunt in Los Angeles, which was miserable. He was often hungry and would be deprived of food. As for Nat, as soon as he had dispensed with me and sent Stanley off to LA, he remarried. He was still alive.

Stanley had not a single question for Nikki or me. About who we were, how I grew up, how I became who I am.

As the sun set, we went for a walk along the winding path to an embankment along the Sudbury River. After, Nikki told me that Leah had pulled her aside while Stanley and I were walking ahead. She pointed at our rear ends. "They are clearly brothers."

Nikki saw these two men, and noticed what Leah saw. Flat butts. Apparently she had never noticed that imperfection in me, but could not forget it after, for which she would occasionally rib me.

Finally, as it grew dark, Stanley and his family left. He was gone.

As I reflected upon this man who said he was my brother, I think the one thing that was most clear to me is that Stanley wasn't very nice. He was curt with his sweet, timid wife, Leah, and paid little attention to his daughter. He had a temper, and used swear words, and was immediately overly familiar with my family. He found me for himself. He told his father, my father, that he had found me, but Nat never wanted to meet me as an adult.

Really, after that weekend, Nikki and I had no interest whatsoever in continuing this as a close relationship. So although we knew Stanley was there, and his kind wife was there, in Vermont, we couldn't envision embracing them as family.

He never asked about how I became me. He died in 2016. We met a few times, much later. I never really knew my brother.

I don't know who I would have been if I had been raised in his family. From the first encounter, I knew that Stanley wasn't someone I could have looked up to in any way, as one might a big brother, as I had looked up to Steven. If I were a Warshaw, would I have lived with that awful Aunt in Los Angeles? Would I have been a wallpaper hanger? Was the difference between being a Warshaw and a Zapol being a wallpaper hanger and being a Harvard professor? Beyond class and education, there was something else—kindness and generosity that I didn't see in Stanley, and must have been missing in Nat, who didn't take on the burden or the joys of raising me or ever reconnecting with me.

My mother, Millie, apparently a kind woman, had been warned against getting pregnant because her first birth with Stanley had been so hard that her doctor felt she might die if she had another baby. Her entire family apparently blamed Nat for getting her pregnant again, with me, which in their minds was putting her at risk. And indeed, she had pre-eclampsia and hypertension, and she died of a hemorrhage in her brain just days after I was born.

After I learned about all of this, my birthday became a complicated day for me, a day of grieving a mother I never knew. I felt that I was responsible for killing my birth mother. This is hard to explain to Nikki, and to my children—who are often full of cheer and love on my birthday. I remember when my children were young, waking me up with breakfast in bed.

"Surprise!" Little four-year old Liza yelled, handing me pancakes with her syrupy fingers, an expectant look on her face. She was followed by her older brother and Nikki, who surrounded me in bed, snuggling close.

"Happy Birthday!" They yelled.

But I just couldn't match their celebratory energy. I got teary. Nikki noticed, of course, because she notices everything. "I'm sure your mother is smiling down at you today."

I forced a smile. Of course, as Nikki has explained, I know a baby can't be responsible for the death of a mother in childbirth. Millie's death wasn't intentional on my part, but my birth caused her death, so I can't help make the leap to fault, to causation.

Apparently, until I was three, my birth father would show up mysteriously at some of my birthdays, but I never knew who he was—he was introduced as a friend of my uncle's. Shortly afterwards he left me behind, remarried, and started a new life in California. He never tried to meet me, and he died in 1997. How could I consider them my family? I just didn't.

Florence sent me a letter. She wrote: "I hoped that this would never happen, that you would never find out. Please don't reject me. I want you to know that I really do love you, as a mother. Please do not tell the children." She was so afraid. Afraid I would know her secret. Afraid I would tell my children, David and Liza, that she wasn't related to them, and that they wouldn't think of her as a grandmother.

Was I angry? Yes, and so was Nikki. I don't think she ever did forgive Florence for concealing the truth, which is understandable, if anything in this mess was understandable. But I also didn't want to get stuck, to get held back. I needed to get back to terra firma.

I called Florence, and I spoke from the heart: "Look, mom. You are just that, my mother. I won't keep this secret from my children and they know already. But I owe you a great deal for raising me, when I wasn't wanted. And I love you, and I always will."

This is how I feel. The woman who wiped your butt and fed you and cleaned you up and did all those things deserves all the allegiance that can be given to her, whether or not she genetically donated half your chromosomes. I am so grateful for her and Ben for taking me in, and for making me who I am. I was sorry I wasn't able to tell Ben the same—he had died by the time I found out what he had done for me.

It wasn't until Florence died in 1999 that I discovered the extent to which my entire family and extended network of friends were in on the secret of my adoption. We buried her in a cemetery in Jamaica, Queens, next to my father, and at the ceremony, I spoke for the first time publicly thanking her for adopting me. I thought I was revealing my secret, but just saying those words thawed the ice around the past, and I could see the expressions on my relatives' faces soften. My cousins confessed to me, one by one, that their parents disclosed the terms of my adoption but forbade them to share it with me.

My pretty cousin Linda, Steven's baby sister, shared how she heard from her parents that in 1942, when I was born, Florence had been trying to have a baby for ten years of marriage, and hadn't been able to conceive. They were on rocky ground as a couple. (Had Ben been unfaithful? I was afraid to ask.)

"Of COURSE we all knew," my cousin Jeff laughed. "Only you didn't know."

It was at the graveside, after I threw dirt on my mother's coffin, that it dawned on me that my entire childhood with Steven, he knew. Now, nearly 60-year-old men, we huddled together and Steven whispered to me that his whole life he wasn't allowed to point out the difference between my mother's crystal blue eyes and birdlike features and my dark brown eyes and bulbous nose, that clearly I didn't look like my parents. He couldn't state the obvious that I had none of my mother's grace (but he did), and that I wasn't actually biologically related to him at all.

I suppose it is funny, and mighty strange, that the pretense was kept up throughout half of my life. After I found out, I had several months of anger and upset about the secrecy around my adoption, and I felt some of the shame that my mother must have felt to keep it secret for so long. But I was in the middle of my career and raising kids. My entire sense of self was built on being Warren Zapol, and although I was shaken, I didn't want it to overthrow my identity. So I suppose I put my feelings into a little box, and compartmentalized it. And then, all of a sudden, when my mother died, it became normal. It became easier to talk about with family members, friends, and even strangers. I began to feel proud of being adopted.

In the summer after she died, I cleaned out my mother's apartment in Miami Beach. I put away her ball gowns, her tchotchkes, and all of my gradebooks that she kept. In a desk drawer, I found her diaries, which I trepidatiously opened. I found the one from 1942, the year I was born to the Warshaws. Two weeks later, Florence wrote an entry: "I'm so happy with Warren I could shout from the rooftops!" My tears began to flow. I wish she had been able to speak about her choice to adopt me without fear of rejection and judgment—most of all from herself. I saw it plainly in her cursive scrawl here. I was loved. I was a lucky kid.

My life was full of people who adopted me, beyond my parents. People who shaped and defined me, who extended great kindnesses and brought me into their homes. I have always been a loyal friend and colleague, and this experience of learning about my adoption later in life has taught me to never take these relationships for granted.

Similarly, Nikki and I always had extra room at our table, to "adopt" people into our family. I always welcomed people into our home and into the warmth of my heart. I helped several Jewish Soviet refugees resettle in Boston, in the 1970s and 80s, when they were persecuted at home. Many of these refugees ate at our table, stayed at my home, and became my best friends, like Aza Raykhtsaum, who is now a violinist in the Boston Symphony Orchestra.

I have become very close with many of my fellows, my students, my mentees, far too many to name, but I hope that they feel as close to me as I do to them, and that they see me as family. Far more than blood, it is the extension of love, care, generosity, and kindness that binds us together.

Do these relationships make me who I am, and do I in turn shape others? What defines us, makes us human, and original thinkers, and risk-takers, and people who think outside the box? How do we get to be the way we are? I may search the whole earth and universe and human body and never know the answer. The world is full of these unsolvable mysteries.

CHAPTER 3

How to Not Get Killed in Antarctica

* * *

Antarctica

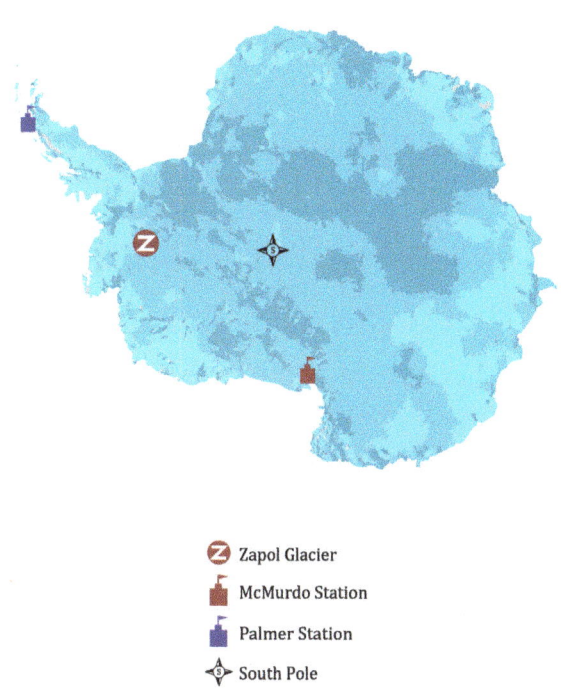

Z Zapol Glacier
McMurdo Station
Palmer Station
South Pole

Map of Antarctica based on USGS and NASA satellite imagery, by Elliot Zapol.

McMurdo Sound

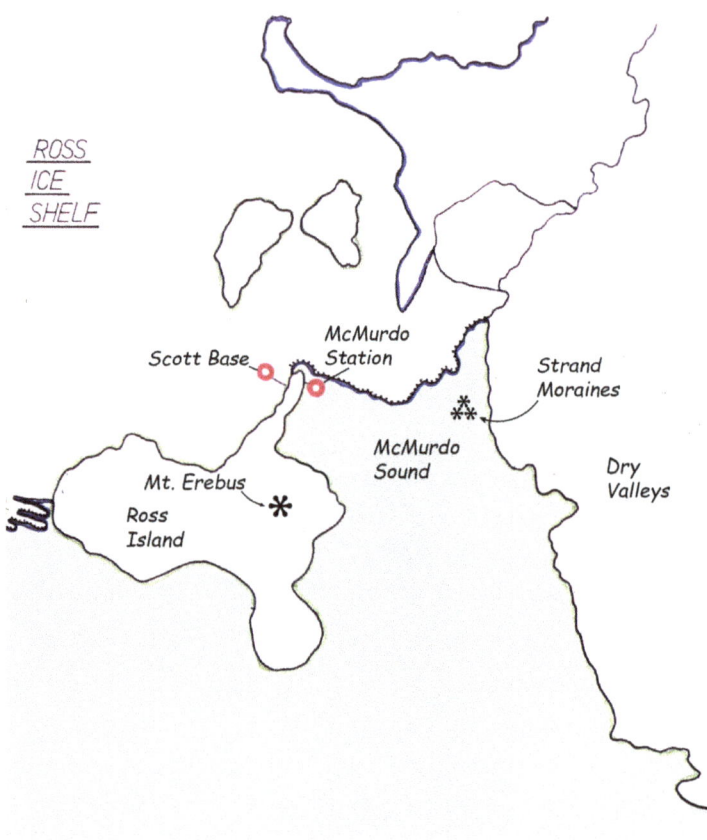

WEDDELL SEALS HAVE BIG BEAUTIFUL eyes and a thick coat of blubber to keep them warm in Antarctic waters. Like all seals, and just like you, they are mammals. They are warm-blooded, and the seal mamas carry their babies inside them, warm and happy in their placenta. So imagine you are going to dive into the icy Antarctic waters. Imagine you are going to dive without a scuba tank. Or let's raise the stakes—imagine you are a pregnant mother and you are going to do this with your baby. If I told you that you and your baby are going to dive 1600 feet to the bottom of McMurdo Sound so you can catch a big meal of a fatty 80-pound Antarctic Cod, how would you prepare for this? Would you start by taking a big breath of air?

I've asked these questions hundreds of times. I've asked them at the Waldorf-Astoria in New York City and on boats to the Galapagos. I've asked world-renowned biochemists. I've asked mechanics in the mess hall at McMurdo Station in actual Antarctica, over a plate of five eggs. (When you are in Antarctica, you have your choice of five or six eggs for breakfast. Mammals without blubber burn a lot of calories in the cold.) I've asked our country's leaders when I was appointed by Presidents Bush and then Obama to the United States Arctic Research Commission. I've asked my kids, my grandkids, students at their schools, and posed these same questions in my lectures to Harvard students and to fellows in my hospital.

With Alaska Senator Lisa Murkowski and Mead Treadwell, Chair of the U.S. Arctic Research Commission, at a USARC meeting, 2009.

I love these questions because we can all relate to them. We all breathe and we all know what it is like to be a diver. But not the kind of diver you are thinking of. Not the kid with goggles, pinching their nose with their fingers, and sinking to the bottom of the swimming pool. A different diver. We all start out as divers. In our mothers, during pregnancy, we are all little seals, holding our breath, and not using our lungs. Until one day we come out into the air and we start to breathe.

But when we're little divers holding our breath in our mamas' bellies, we don't hold our breath, actually, do we? We don't hold oxygen in the air in our lungs because we're not breathing air in the womb. Our lungs aren't working yet to exchange oxygen. In fact, we hold oxygen in our blood. Our hemoglobin binds to the oxygen and holds it until we need it in our muscle or in our brain. Amazingly, our hemoglobin while we're in the womb is stronger. This is fetal hemoglobin, and it is better at binding to oxygen than our mother's adult hemoglobin so that we can pull enough out of the placenta and umbilical cord. We need to steal our oxygen from our mamas because we're not breathing air.

Seals do similar tricks, and that is why I went to Antarctica.

To be clear, I was driven first by the prospect of adventure, coupled with the possibility of doing science in an extreme environment where fierce storms, four months of rock-solid ice and dark winter were the backdrop for extreme adaptation. I had no experience in such an environment. Many of our greatest explorers and adventurers are from Norway or Alaska or Russia. I'm from Brooklyn. Brighton Beach is about the farthest most Brooklynites get (that's still in Brooklyn, for readers from other boroughs and beyond). But I'm a scientist and risk taking is in my nature, you know. Risk and reward are somehow tied together. If you take a great risk, and win, you could have a great reward. When we went to Antarctica we took risks, both in our own lives and in the science we did. But we pulled hard, and if you bring together a team that is spectacular, and if that team pulls together, and the weather isn't too terrible, you're likely to succeed. Somebody will get the right idea while you are there, and even in a short period of time, you will advance human knowledge. Every year isn't great but often I found, with my trips to Antarctica, the first year on a new project was just about figuring out where the technical problems were, and only in the second year were we able to conquer the problems proposed by our study and bring home the bacon.

I was spurred on by an intriguing question posed by Dr. Jesper Qvist, a smart, spry Dane who worked with me in the operating room at MGH. Remember, we're anesthesiologists, so much of the time our patients are asleep and maybe the surgeons are doing their work, and we're monitoring the machines, and nothing happens on our end for a few hours, so we have some time on our hands to trade jokes, or get to talking science.

In 1973, Jesper and I were in the operating room during one such lull in the action, and he said: "Warren, we chill people and we know there is a lot of benefit to that. Maybe there is some part of that we can understand from animals in cold places?"

At that time, there was a lot of interest in the pH of blood. pH is controlled by lungs and kidneys primarily and must be maintained for good health. A normal pH measurement is around 7.4. Buildup of carbon dioxide (CO_2) as indicated by a decrease in pH, makes blood acidic. We decrease CO_2 by exhaling it. People were wondering what we should do to control pH of human blood when it was cooled, like in pulmonary bypass for heart surgery, down to 20°C, versus 37°C, which is the temperature you and I maintain when we're healthy. Patients' pH levels shift upwards as we cool the blood, not a desired result. We were measuring blood pH all the time in the operating room to make sure it wasn't getting too far out of balance.

We found a book, *Biology of the Antarctic Seas*, by the Head of Biology at National Science Foundation's (NSF) Polar Programs, George Llano, a Cuban-American, Harvard-trained scientist. In it he described fish that live in water below freezing and seals that dive for over an hour. It was like a veil was lifted and I saw a whole world of biological adaptation to the same cold temperature we were making in the operating room. With that one book I was hooked on extreme biology and my life, my science, and my whole family would never be the same.

We wrote to George, asked for a small grant to cover airfare to Antarctica that fall, and a couple hundred dollars of supplies. He and his wife came and visited me and Jesper and our families out in Concord.

We had a community garden in our neighborhood, where we had a plot for many years growing tomatoes, strawberries, asparagus and some herbs. Nikki and I walked to the field that morning with Jesper to gather some strawberries for the visit, talking about how hard it was to imagine Antarctica, how strange it was to think of a whole continent covered in ice. We walked back and George and his wife arrived. We sat outside, a warm spring day, looking up at the pine trees and talking about science.

"Maybe fish are like us," I said to George and Jesper. "If they are like us, they should also have a higher pH at lower temperatures. We've been working in the lab measuring pH for years. We definitely can measure it in Antarctic fish and find out."

An avuncular fellow, George smiled, and surveying the red strawberries on the table and the green pines above us said, "You've never been anywhere like the white continent of Antarctica. Go with someone who knows what they are doing so you don't get hurt."

George and his wife left, wishing us well and thanking us, and miraculously found us a place in a grant with Art Devries, an Old Antarctic Explorer (OAE) who studies Antarctic fish, and who I thought resembled Poseidon, the first thick Antarctic beard I would encounter. Jesper went down early in October and worked with Art in the aquarium at McMurdo, where they figured out how to put arterial lines in healthy fish in the tank, and then drew blood samples and set up a blood-gas analyzer in the laboratory. The science was challenging—we always had to plan ahead to bring everything we needed, but the biological laboratory had a backlog of 25 years of scientific expeditions, so we could usually improvise something we might need but forgot to ship down to the ice, like a broom handle or some duct tape. But that came later. I love that part of extreme science—making do with what you have and trying to get useful data out of the experience anyway! I joined Jesper towards the end of October to do some biochemistry, and we were like two peas in a pod, measuring lactates and pyruvates, blood metabolites, trying to see how different these creatures living in freezing waters are from us.

My long journey began by leaving the brilliant fall color that surrounded our hospital in Boston. In early October, I flew into the smog of LA where we'd be measured for the red suits we'd need to be visible in the snow, so that

1976 research team. Standing, from left: Jesper Qvist, Mont Liggins, Peter Hochachka, Tommy Wonders. Sitting: Paul Wankowicz, Mike Snider, me, Bob Schneider.

if we were lost in the ice a helicopter could spot us. Then we'd stop over in Honolulu, say goodbye to the sand and the tropics, and drop south to lush, rainy New Zealand. From New Zealand we'd be in a holding pattern for clearance to proceed further. We'd walk in the gardens of Christchurch, gathering last supplies and toasting farewell to green things, and then, once notified that weather conditions were favorable, we'd depart on a midnight flight in order to catch the morning calm and light in McMurdo.

I walked onto the plane and there weren't any tickets, no assigned seats. In fact, there were no seats! The C-130 cargo plane seats were just nylon straps. About as comfortable as sitting on a seatbelt. We'd lie down in the makeshift seats or on top of cargo and after a few hours, try to get a glimpse of the continent out of the window. Sometimes the Navy pilots would invite me up with my Super 8 movie camera to shoot ice fields or the coastline of the continent and talk with them, sharing stories of our adventures. It was early in the summer season, so we had just a little darkness during the flight. If the weather was bad, we'd have to turn around after four hours and try again another day. If we were lucky and the weather held, we finally landed on bright white, six-foot-thick ice outside of McMurdo Station in Antarctica, with the Eagles blasting "Hotel California" out of the cockpit: "You can check out any time you like, but you can never leave."

The red parka I had been issued by the Navy had a hood with fur lining. When cinched at the neck, it formed a long tube in front of my face to keep the elements out and my breath warm. When I stepped off of the plane, I had my hood off and I was blasted by the ice cold wind and immediately the hairs inside my nostril froze. I coughed from the tightness in my chest and flipped up my hood. That made it hard to see out, so I stumbled down the stairway and onto Jesper.

As he steadied me on the ice, Jesper said, "Welcome Warren! We're learning how not to get killed in Antarctica."

He threw my bag into a big wheeled vehicle, the Terra Nova bus. Terra Nova looked like a Star Wars contraption with small windows. When I climbed in it was filled with people in red puffy suits like mine and beards of various lengths. Very few shaven faces, and even fewer women. We trundled into McMurdo "town"—an odd collection of buildings strewn across a volcanic landscape with the mess hall that offers five or six eggs for breakfast, the biolab, aquarium and mechanics shops, carpenter's shops, and a point with a hut where Captain Robert Falcon Scott set up base on his ill-fated path to Pole and back in 1911. Towering over the town is an oddly shaped cinder cone named Observation Hill, affectionately called "Ob Hill."

Ob Hill has a memorial cross on top of it inscribed with the last line of Alfred Tennyson's poem "Ulysses": "To strive, to seek, to find, and not to yield." These weighty words commemorate Scott. On a later trip, I encountered another tragedy on Ob Hill.

That time when I got off the plane, Tommy Wonders, a marine corpsman who I met in Vietnam and who ran my lab at MGH, came up to greet me wearing his hood right over his face, so I could only see one eye. A towering and rotund presence, he was a formidable Cyclops.

Tommy said, "Hey, wonderful, Warren, I'm glad you're here!" Then he shook my hand and said, "See you back at the base!"

I climbed into the Terra Nova bus and Tommy headed for another vehicle. I dropped my bags in the dorm, not coincidentally named "Hotel California," after the song that had been blasting from the cockpit on my first trip, reminding us all of the inescapable nature of Antarctica, and the irony of being literally the polar opposite of California. I walked across to the biolab where we brushed off the snow and ice in the vestibule and Tommy opened his hood up, revealing a big black eye. He had fresh sutures running down the front of his face, Frankenstein style. I preferred Cyclops.

"Tommy, what happened?" I stammered.

He winced and said, "Well, long story. I was afraid you were going to send me back on the plane, so I waited 'til it took off for Christchurch for me to take off my hood."

I said, "Ok, it's gone now. We're here in Antarctica trying not to get killed. What happened?"

He replied, "Well, I've had some extra time on my hands here…"

I nodded. I had sent him down a month early to build our operating room, and set up the camp on the annual frozen ice surface. He was always fearless in his commitment to the lab, and I could imagine any number of ways he could have done this to his face, wondering whether he'd fallen in a crevasse, or if some heavy equipment or dynamite was involved.

Tommy said, "I was here at McMurdo gathering supplies for the camp, and the hospital corpsman and I decided we would go sledding one sunny day. The weather was perfect, just below freezing. So we took the ambulance."

Tommy had been a hospital corpsman in Vietnam and he could befriend anyone with his beaming bearded smile and twinkling eyes. He was a friendly giant. So he'd made a friend, which wasn't all that surprising. Well, the two of them took the ambulance up the side of the glacier on Ob Hill, parked, and got out banana sleds. Banana sleds are plastic—they look like banana peels; They're basically hard plastic banana peels. They have no steering capacity. They have a place to hold on, a rope for dragging it behind you when you walk back up, but they have no guides, no glides, no skids. They're just a smooth, plastic bottom you sit on and slide. Tommy would take off, go down aways, and then the corpsman would drive down in the ambulance, pick him up, and they'd drive back up, and then the corpsman would go down. They'd take turns, one of them, then the other one, hurtling down Ob Hill. "To strive, to seek, to find, and not to yield."

They did it a bunch of times, and it was fine, a lot of fun, and they got up to 25 miles an hour. So Tommy and the corpsman got into a bit of one-up-manship which is something that happens in sledding, where you go higher, faster and more out of control each time. My kids used to do this on the hill behind our house, and it resulted in some scars still visible today.

Tommy explained succinctly, "We were going at about thirty miles an hour on the last trip down when my face hit the rocks, or the ice, or ...well, I can't remember quite what I hit."

I thought about this. From the looks of it, his face hit ice—there was no volcanic dust in the scrapes, but I would need to look under the bandages to verify.

"Fortunately there was an ambulance and a medical professional on hand, so the corpsman threw me in the back and sewed me up at McMurdo's general hospital."

He finished up with an apology, "I'm so sorry Warren, I don't know what I was thinking, except that it was really fun while it lasted. I know I'm not here to sled, and I'm committed to making the science a great success. I'm just thankful you didn't send me back to Boston."

I paused for a moment as I considered sending him back to Boston.

"WHAT THE HELL WERE YOU THINKING?" I looked him straight in the eye and asked. I had known Tommy since we met in a war zone. We'd literally saved each other's lives, saved countless other lives, and I was scared of what could have happened to him if this stupidity had caused worse damage to his noggin.

It was a few hours before I started seeing just how funny it was, and that night Tommy had the cooks in the mess hall doubled over with laughter at his tale of flying down Ob Hill at just-below-break-neck speed.

I also enjoyed great times of levity. We put on skis and slid around on the ice together. Once I dragged Tommy behind a tracked vehicle on skis in a frozen water skiing adventure. One day we pretended we were on Miami Beach, one of the team took off all his clothes and went sunbathing. It was minus ten or something! Although if the wind stopped, you could

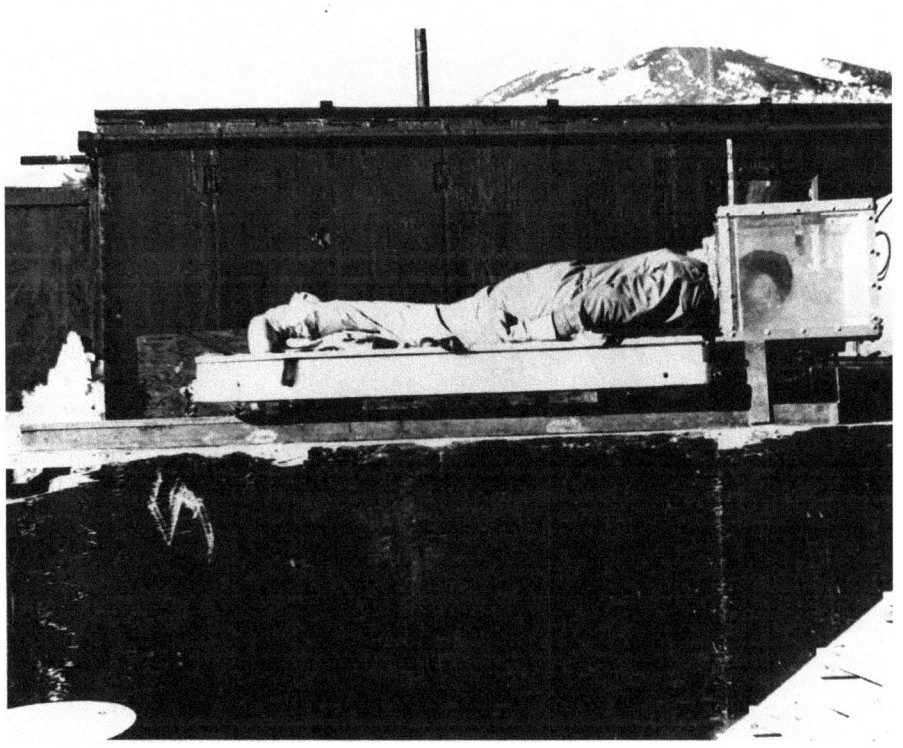

Tommy Wonders demonstrated how to place a seal's head in water to simulate diving.

get a lot of solar radiation and feel quite warm. I wasn't naked—forget that thought—but we'd all be in T-shirts when it got that warm. Some Emperor penguins lost from a nearby rookery wandered by and we wondered what they thought of us.

For Halloween there were costumes and bands in McMurdo. People were very invested in partying, letting off some steam. In general, I spent five minutes at the party. On Thanksgiving Tommy roasted a turkey in a glassware oven, and one year we made duck a l'orange with a heat lamp. These were all great times, we'd have a good laugh and share stories from home.

In our first year working on fish in the biolab, Jesper and I met Don Siniff of the University of Minnesota, an ecologist who had worked on seals. I spent a day with him counting seals by Little Razorback, an island in McMurdo Sound. That was a pivotal experience for me. I came back to McMurdo and said to Jesper, "Fish are great. Seals are better." The logic was clear to me, and I explained, "We're people doctors. Seals are closer to people. Fish are fish, and, you know, seals came along thirty-five million years ago and fish are, I don't know, a billion? A half billion?"

He laughed. "Okay, you're not a fish biologist. It's 375 million years back to our last common ancestor with fishes, so seals are closer to us?"

Yes. We wrapped up the fish studies and found that at -1.9°C, Antarctic cod have a high pH (8.2 - 8.3) as we had predicted, much higher than normal human pH of 7.4. Then, propelled by my new fascination, we submitted our next grant to George at the NSF on seals, the longest, deepest diving mammal in the world. On it, we added two new members to our expedition team: Peter Hochachka, from Vancouver, and Graham C. Liggins ("Mont"—later "Sir Mont"), from New Zealand.

Peter Hochachka joined us thanks to Bob Schneider, my dear friend and fellow anesthesiologist. One day, Schneider, who was also taking advantage of down time in the operating room reading a copy of *Science*, looked at me and said, "Gee Warren, there's this guy Peter Hochachka who writes about hypoxia and low oxygen levels and all this other stuff like evolution; I wonder if he would go with us to McMurdo to study metabolism and pulmonary circulation of seals?"

Bob called Peter in Hawaii, where he was studying giant tuna and Peter didn't miss a beat.

"Sure, I'd love to work with you! Can you get me some blood and muscle samples?"

Peter, a tall, wiry, beaming Canadian who always looked like he had just rolled off a ski slope, was a kinetic ball of energy, constantly bouncing around and telling fantastic, zany stories. Most of these were about the animals he studied, in extremes—giant tuna swimming in tanks where he'd use dynamite to sample their muscle, hummingbirds flying in air chambers that were frozen solid with liquid nitrogen mid-flight, Lungfish breathing air while peeking out of the Amazonian puddles, and Andean soccer players running on treadmills (also conscripted to play with his kids' soccer teams in Vancouver BC), for example. He was irreverent, and just the kind of fearless thinker we needed on the team in order to inject new thinking.

Weddell seals, above ice, are cute but slow.

I first met Mont Liggins in Buenos Aires, Argentina, in 1974, when I gave a lecture on ECMO and the artificial placenta at the World Congress of Obstetrics and Gynecology. I already knew of him, a visionary doing research on fetal lungs. A gentle, gnome-like figure, I was immediately drawn to him. As Mont and I chatted over a scotch one night, he confessed: "My grandfather was a provisioner to Shackleton and Scott in Christchurch, and I've always wanted to go to the Antarctic."

Mont would learn during our half-dozen trips to the ice together that he preferred his Glenlivet with glacial ice, which was clear and hard, and gave off bubbles of gas frozen in time. I would learn that I don't like scotch unless I'm drinking it with a cigar celebrating a marriage or a birth, and even then I prefer port. No matter this difference in taste, we became fast friends, always feeding off and into one another's insights and ideas. Our work, like that of all serious scientists, contributed to a slowly woven tapestry of incremental collaboration.

Mont had a knack for making things simple and understandable for those of us who weren't as familiar as he with reproduction. I remember his instructions on our first trip out looking for seals to study:

"Look for the pregnant mothers," he said.

"How do you know if it is a male or female?" I asked.

Mont smiled on one side of his face, "Look for his pecker!"

Mont's great contribution to humanity was that he figured out that if you give a shot of a hormone, cortisol, you can save babies that are born prematurely. Their lungs start to function and the babies can breathe. For that discovery, he was knighted by Queen Elizabeth II and became Sir Mont. This work has inspired the Bill and Melinda Gates Foundation's work to make cortisol available worldwide.

With Peter and Mont on board, we had assembled an "A Team." I always was a big believer in team building, and there were many team lessons to learn from the adventurous explorers of Antarctica. I collected their books and stories for years. When I looked at the map of Antarctica, I imagined the historical paths to the South Pole. I envisioned the places in the ice where ships were crushed and sent to the bottom. I remembered Ernest Shackleton's boat, Endurance, that sank to the bottom of the ocean while he saved his entire team. Shackleton is my greatest hero. His and Captain Robert Falcon Scott's huts remain intact 100 years after they were used to support missions to watch penguins in the winter or gather geological samples across the continent. My work was built on the shoulders of these giants. When we went out to set up our field camp, our team slept in tents named "Scott tents," less than 200 miles from where he died in one. We drive the same Tucker tracked vehicles that carried Sir Vivian Fuchs and Edmund Hillary to the Pole in 1958, who departed from and returned to none other than… Scott Base. We're only able to do our science because of the teams and explorers who blazed the trails and built the knowledge base for us.

One of the most beautiful sights I've seen in all my travels across the world was watching a dog team crossing the ice in Antarctica pulling a Nansen sled. It looks like a Fred Machetanz print from Alaska, or a remote rendition of a Hokusai print of Mt Fuji. It echoes the explorers like Shackleton who used them to carry their supplies on their expeditions. The Brits and Kiwis had dogs at Rothera and at Scott Base, so we'd see them in the 1970s and 1980s and even the early 1990s crossing the ice heading out for a mission. We'd go over and photograph them and help to feed them... more on that later. The history was very much alive all around you in McMurdo Sound. I could ask for the key to Scott or Shackleton's huts from the Reverend in McMurdo. Why he had it, I have no idea, but I could let myself in and look around. There was no real care of the huts in those days. It was extraordinary and I couldn't help but be interested in the history—to dig into why this place was named this or that... and the answer often was that some seaman had drowned there, or some explorer had died. It was eerie to see crackers in tins waiting for Scott to return from his expedition to the Pole. On a warm day, I could still smell the blubber of seals that Shackleton burned for fuel at Cape Royds.

Antarctica isn't just filled with scientists and engineers—there are also wonderful artists and writers who make it down there. They portray Antarctica today for the generations of people who dream of the southernmost continent but will never make the journey. My great friend Meredith Hooper, a gifted writer of children's books and histories, has spent many years bringing to life the stories of the explorers and their animals, of today's scientists and support teams. She was one of the first to relate the story of scientists at Palmer Station who collected data on climate change by studying the disappearing habitats of the Adelie penguins. Galen Rowell, one of the great Yosemite climbers and photographers, arrived at our fish hut on an expedition, where he caught some stunning pictures of seals in the ice. I'll never forget his surprise when a seal surfaced in the hut while he set up his cameras and blew fish breath in his face.

Early on, we relied on Don Siniff to go with us to find our first seals and bring them back to the biolab for our experiments. We would inject radioactive substances and measure blood flow to various organs. We were interested in how the seal chose to distribute precious oxygen during a long breath hold, so we'd set up a "simulated dive." Like a seal bobbing for apples, this dive was created by putting the seal on a wooden plank and then tilting up on the tail-side and sliding its head into a water bath. We'd keep them there for a short dive of 15-20 minutes, something you or I would not be able to tolerate, but for these diving beauties it was no problem. Then we'd let them breathe room air and recover and have them dive a second time for 30-50 minutes.

We would be taking samples as we went, but in order to harvest the organs, we had to sacrifice the animal, which we would carefully do with a procedure designed with the supervision of veterinary experts back at MGH and through the skilled administration of anesthetics.

We had received permission from the National Marine Fisheries Service and National Oceanographic and Atmospheric Administration to take 20 seals. There were 1,000 in that colony at Little Razorback, and dozens of colonies all over McMurdo Sound. We didn't influence their population at all, and the ecologists like Don were studying the numbers to make sure we weren't causing harm to the colony. But we had to figure out what do to do with a radioactive 1200 pound seal. Once we had taken out the organs and measured their radioactivity and their size and weight, we'd gut the seal, throw away the guts, and take it over to Scott Base. Sled dogs at Scott Base ate a diet of seal meat. You got the picture? So we put the seal in an ice crack and planted a flag over it so we'd know where to find it. The bodies would freeze solid, and the isotopes would decay. After a safe period of time the dog feeders would take a buzz saw, cut off cubes, throw them into the microwave, and feed it to the dogs. It was a perfect ecological use of seal meat. The next year Mont, Peter, Bob, Jesper and I walked over to Scott Base for a drink at the bar one sunny evening, and we stopped to see those dogs—they barked and seemed happy and well fed.

We liked walking over to Scott Base because it was small and intimate. We never loved McMurdo Station, because it was so bureaucratic, with rules and courses to take which got worse with the advent of the internet and new ways for Washington to control us in Antarctica. When we were out on the sea ice at our camp we were alone and could do what we needed to do to get the science done. If we checked in by radio with McMurdo Operations at eight in the morning, they would know we were alive and they wouldn't worry about us until the next day. So, being on the ice was always the most fun. No one watched. We had skidoos, track masters, seals, and we'd try to solve our problems, get our blood samples, mount our devices, make our measurements, and what happened was between us and the seal's physiology. Of course we'd tell people back in McMurdo and in Washington if something was not going according to plan, and we always played by the rules as best we could, but the time out on the ice was a time to be creative, and there already were plenty of constraints on that creativity just based on the limitations of operating in a remote environment where the biggest store was a warehouse 25 miles across the ice.

Being on the ice for two months was our opportunity to learn as much as we could—and that is just what we did! By working around the clock and tinkering with our electronics and lab studies and seals we came up with articles for *Scientific American, Science, Nature,* and a stack of academic journals, all around answering the question: How do those seals manage to dive? Ultimately the insights came from really paying attention, observing and learning about the animals in their native habitat. Louis Pasteur said, and I often repeat for my students, "In the fields of observation, chance favors only the prepared mind."

We published some wonderful findings about how the dive reflex worked, but in the early years, one piece of feedback in the peer reviews bothered us

Weddell seals, underwater, are gifted divers able to dive
1600 feet to the bottom of McMurdo Sound.

because it undermined all the findings. We were sitting in the biolab and the wind was howling through the cracks when Peter summed it up nicely: "They think we are eliciting a 'fright reflex' by putting a seal's head into a bucket!" He said.

In other words our "simulated" dive conditions could have been simply generating a response to an artificial environment where the seal couldn't control the timing of its next breath. We didn't believe this was the case, but we had to do something radically different if we were going to get over that hurdle. After much debate we settled on monitoring seals in isolated holes. The concept was straightforward: We could set up our laboratory in the middle of an ice-sheet, placed strategically over a hole to which a seal would have to return after a dive.

Our laboratory thus would move from McMurdo Station out onto the sea ice, and home for us would become a fish hut—literally an orange plywood box that rested on sled runners, with a stove fueled by an oil barrel hanging

The team somehow corralled seals into a sledge and dragged it across the ice to the isolated laboratory near the Strand Moraines.

on the outside. We could pull the hut-sled out onto the ice with a tracked vehicle and place it where we wanted it. In early years we would use dynamite to blow the hole in the ice. But later we'd get help from a crew of heavily bearded Alaskan mechanics and equipment operators who would drive out with us those 25 miles to a point near the Strand Moraines, across McMurdo Sound. They would drill a hole in the sea ice with a massive ice-auger on skids. Imagine a drill press with a 6 foot wide bit that can go through 10 feet of solid ice in about 10 minutes before it plunges straight down to the ocean. Sploosh! The water would come gushing up onto the surface of the ice, and we'd all cheer. Then we'd drag the fish hut over the hole, light the stove to keep us tolerably warm and the hole from freezing over. Refueling the stove required weekly delivery from McMurdo by more bearded Alaskans, who would also drop off a few weeks of rations as well in case we got trapped there in a storm.

We glued computers onto the seals' backs so that when they returned to the hole we could gather blood samples and real time oxygen saturation in

The Zapol laboratory, placed over a hole drilled through solid sea ice.

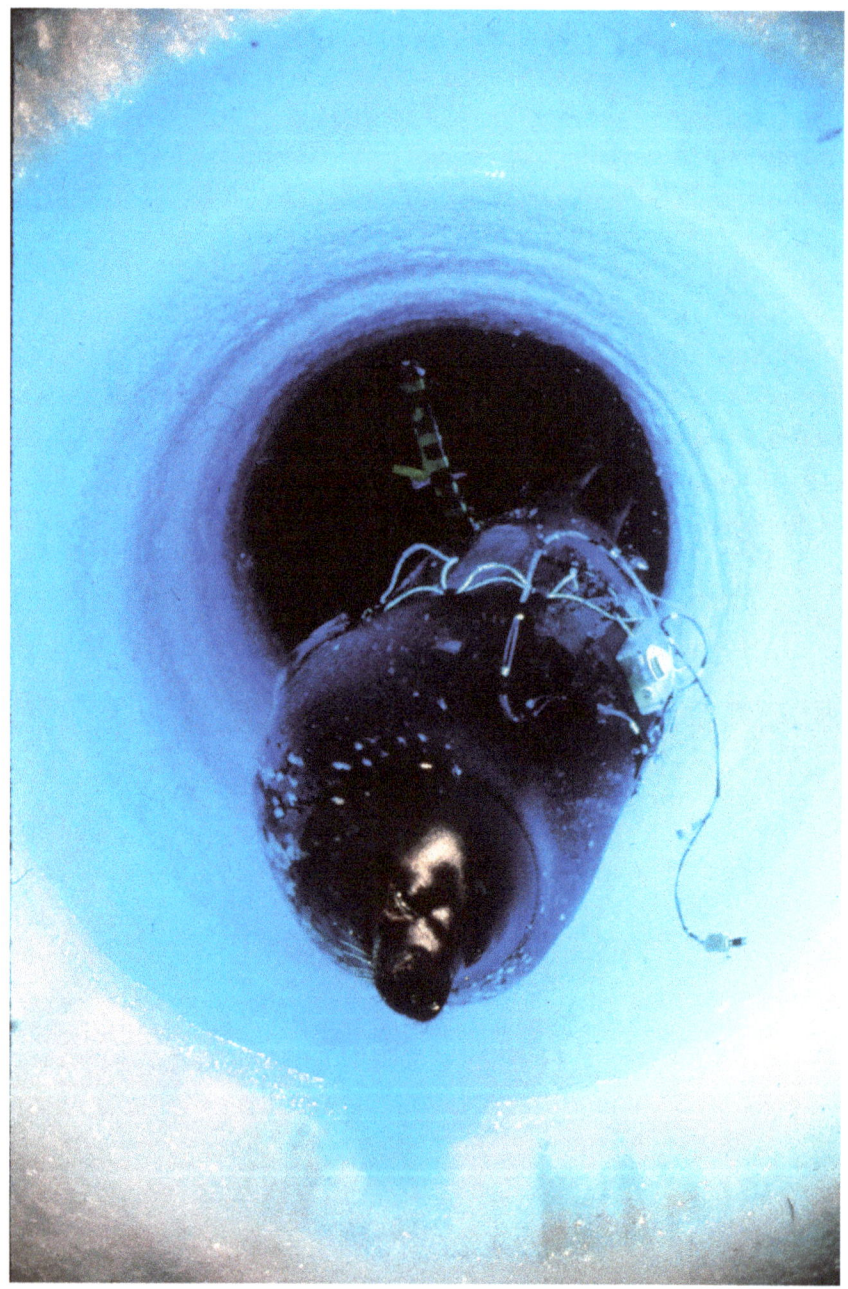

A Weddell seal, returning to the surface in the isolated lab.
The seal has a computer on its back and a fiber optic cable, ready to download data.

muscle and blood, and other data profiling the dives in real time. That engineering feat took many years to perfect, but ultimately paved the way for a complete understanding of how seals dive to the bottom of McMurdo Sound to catch those fatty Antarctic fish. We built custom computers engineered so when the seal surfaced we'd connect a fiber-optic line to download 64 kilobytes of memory, impressive for 1984. Over time we thought we'd be able to use satellites to transmit the information in the wild, but the low earth orbit satellites were rarely overhead where we were—800 miles from the South Pole. We had about a bible worth of information we wanted to transmit, and only a sentence would get up to the satellite in each pass. That never worked for us, so we kept the fiber-optic line.

If you ever meet a Weddell seal surfacing in a hole in the ice while you are in a fish hut over McMurdo Sound, the first thing you'll notice while you wipe the spray off your glasses is that it is noisy. It isn't obvious viewing from the surface that the seal is some 10 feet long and 1200 pounds. You will hear the seal breathing deeply for a few minutes, its nostrils pinching closed between breaths: pushing out air, spouting more noisy plumes of warm humid air into the room. You can watch the shiny creatures bob up and down as they breathe, becoming more buoyant as they pull in air, and then dropping down as they let it back out into the room. It's hypnotic, and if you've been up for 24 hours doing experiments it can lull you into day dreams… until "Pffft!" a plume of spray will wake you back into consciousness. As the seal gets ready to go down, you will be able to tell if it will be gone for a long, deep dive or a shorter dive by the strength of its last exhalation. If it is going to go long, and presumably deep, it exhales deeply. Its forehead, eyes and nostrils dip below the crystal clear water's surface (in that order), and bubbles begin to rise from its nose. Its lungs and entire body are squeezed by the pressure that builds with each meter it descends into the dark ocean until at about 40 meters, we discovered, the lungs collapse.

Now, if you've ever floated in a pool like a kid with goggles, you know if you exhale and blow out bubbles, you'll sink like a stone. Weddell seals are fat mammals, much fatter than any human. They have a thick layer of blubber that protects them from the cold, and that also makes them buoyant, the most graceful creatures under the sea ice. We started to catch a glimpse of their swimming trajectory from the graphs we would generate from the computer data we'd download when they surfaced in the hut. Later Jerry Kooyman and others would actually put video cameras on the seals and entertain us with "critter cams," which made it possible to watch seals fishing and bringing their catch back up to the surface. The professional videos and pictures of them diving under our fish huts that have been shot over the years are some of the most beautiful images I ever saw. They look like ballet dancers as they pilot themselves around the blue and white landscape under the ice. I had a few of these pictures pinned to the cork board behind my computer in my lab back in Boston.

If the lungs collapse, and they don't keep a breath of air for the dive, how do they do it? One of the great adaptations the seal has made to its aquatic life is to store oxygen in its blood when it dives, and not in its lungs. How did it manage to do that?

1992 research team. From left: Bob Schneider, David Zapol, Greg Guyton, Mont Liggins, Kevin Stanek, me, Peter Hochachka, Bill Hurford.

Well, in 1992 my son David and I were together on the ice, and we started to actually look at this before and after dives. David was twenty years old and took a year off of college, worked in the lab and even made the sacrifice of having his wisdom teeth pulled in order to get permission to go. David was generally the one holding the anesthetic over the seal's nose when we'd glue the computer on her back. The folks in McMurdo referred to him as "YZ" or Young Zapol, and I was somewhat taken aback at the prospect of being, at 50, "OZ." Fortunately, Mont came up with a (somewhat) better moniker and I was dubbed "XYZ."

We had a new tool to look inside the seals: Mont had brought one of his ultrasound machines, similar to one that is used to look at a baby in a womb, but portable so we could fly down to Antarctica and drive across the sea ice to our hut, and even go into a seal colony with the ultrasound in a backpack!

I was the last team member to arrive that year as I was hard at work at MGH trying to find a licensee for our nitric oxide patents. Mont met me in town when I landed and we had to drive out to the field camp. In order to drive in Antarctica we had to step up onto the giant tracks of a vehicle and then into the cabin. We had white-out conditions as I drove across the sea ice the first time that year. The noise of the engine driving the tracks across the ice was deafening. The cabin is a box with a heater. You have a gas pedal and two sticks which control the breaks, one for each track. Pull two to stop, one to turn. In the back it's just two rows of seats and little windows. The heat blasting and the tracks rumbling, it is a good time to catch up on the news.

Driving around the ice makes it sound deceptively easy to be in Antarctica, and at that point, early in the season, it was still pretty solid ice. Still, we had to be cleared by McMurdo's operations chief "MacTown Ops" in order to go out onto the ice in regular operations.

Leaving town, you would call MacTown Ops and you had to say how many "souls on board," and then off you'd go like ghosts into the white. The wind was blowing snow across the ice that day, and we flipped the hatch over the passenger seat and Mont stuck his head out so he could see over the snow and guide me along the flags that David had put out every 100 feet when he had set up the camp a few weeks earlier, marking a road all the way across the Sound.

Mont told me stories as I drove, and talking loudly over the engine, re-counted the night he had spent huddled in the back of the tracked vehicle during a blizzard, trying to walk around, bracing himself against the cracks in the doors. Mont became hypothermic, and Jesper had made him get up and walk around to stay warm. When the storm broke for a few hours the next day, they backed off the ice tongue near the Razorbacks, and thankfully made it back to base, where the station master gave them a talking to. I was shocked that they had left on the trip without a radio, but mostly I was scared for them and couldn't imagine passing a night like that in a vehicle on the side of a glacier.

I thought for a few minutes, rumbling through the white. Then I shouted back the story I had heard about a graduate student who fell through in the ice not far from that same spot. Where the tongue of the glacier which comes off Mount Erebus and falls into the sea, the ice freezes but provides dicey support as any crack can move in a storm and work open. We were all trained that if you feel the vehicle go through an ice crack you have to get out as soon as you can. Everybody out, the doors fly open. But in a tragic moment, the vehicle hung up, the graduate student went back for his camera, and that was the end. It is up to 500 meters deep in the sound, 1600 feet down. Mont told me he had nightmares about sinking into those icy waters.

To distract ourselves, we turned on the radio and listened to MacTown Ops—monitoring the comings and goings of the other research groups, until we grew bored, and turned to a discussion of our scientific program and the work at hand.

After two hours we arrived at camp, 25 miles from McMurdo, but invis-ible until we were upon it, what with the blowing snow. Mont jumped out, ready to send pregnant seals off with computers glued to their backs, and wires on the fetuses to measure heart rate, depth, velocity and take samples. Despite the journey I was also eager to join the tinkering because there was invariably something that we hadn't planned for.

This time it was the ultrasound.

"How are we going to get the ultrasound probe onto the belly of the seal while it is awake, down in the hole, without getting bitten?" asked David.

"Tie it onto a broomstick!" said Mont.

We didn't have a broomstick, but I'd grabbed a telescoping pole for mounting my radio antenna before I left for the ice.

"Hey Dad, can we use your antenna pole?" asked David.

"Sure," I said, "But if the seal bites it, we might not be able to call home on Sunday."

Tying the ultrasound onto a pole, squirting ultrasound gel onto the probe and wrapping it in a long plastic bag to keep the electronics from frying in the seawater, we were able to submerge it into the ice water next to the seal's belly and gently press it against its side to peer inside. It's a little like petting an exotic zoo animal—they are in as much awe of us as we are of them. The seal's big eyes blink and peer curiously at the pole. The first few times that we tried this, we moved too quickly and the animal would drift back down the hole, then come up again, but eventually we coaxed her into letting us see. Weddells have no land predators, so people are generally viewed by seals, I like to imagine, as big penguins. Fortunately she didn't bite the pole either, and we were able to call home that Sunday.

The ultrasound images were fabulous! We were looking to see if we could actually see the spleen filling back up in real time. Click! Click! We'd take pictures of the ultrasound images and measure them once she was back down on another dive. Calculating the width and the length of the spleen from our pictures, we were able to derive volume. What we discovered was a dynamic process that unfolded before us in black and white on the ultrasound screen. The spleen of a diving Weddell seal is used like a scuba tank, except instead of storing gas, it retains oxygen bound to red blood cells. Red blood cells bind oxygen to hemoglobin at the surface when she is making those puffing sounds—before sinking down in the ocean again. While she is at the surface, the seal's spleen fills up, and as she dives it squeezes in a controlled release, providing oxygen for the seal to be able to swim, think and eat during a long dive. It's brilliant and beautiful.

While the seal is diving, we're stuck on the surface of the ice, in a hut. Sometimes we'll be processing the images or other samples from a prior dive, or talking over an experiment we're planning. Sometimes we'll be studying maps and trying to figure out where the seal colonies are this year.

Before the seals make their final escape from our hole they usually hang around for 24-48 hours, returning to our hole a few times, allowing us to download data from the computers and most excitingly, observe their return, which gives another clue as to their extraordinary diving adaptation. As they come back up, they look like rockets launching from below the earth. They emerge first with pointy nose and big shining eyes, and then the big barrel chest and flippers on the side and back. Sometimes they'll rotate as they rise up, and it feels like they are showing off. When they surface, they exhale first a quick puff, and then suck in quickly to get fresh air with its essential oxygen into their lungs quickly. Oh sweet, noisy, air. Fish breath. The hut smells like a fish market.

David wasn't there one evening when we were working on a seal. Not being a person who's very observant of familial things, non-scientific familial things, I had to ask.

"Bob, where's David?"

Bob said, "I don't know, he's around somewhere."

I pondered this as we laid out the computer and the glue and the bag of saline that would be tied into the pump to take the samples as the seal went diving. Eventually I said, "He's not here, and we have to start this anesthetic. Where is he?"

Bob finally came out with it. "He's out on a date."

And I said, "A date? With a penguin?"

Bob said "No," and smiled a big grin, as Bob often does.

It took me another moment to put it all together. He had met Diana Laird, who was working with Art DeVries, the Poseidon-like fish biologist who was our teacher on our first trip, twenty years earlier. Diana's father, Bob Laird, had worked in Antarctica during the winter of 1962-1963, and, aware of the same restlessness in his daughter as I saw in David, made the connections that brought her to McMurdo as a lab tech. Diana was working on fish antifreeze with Art, and David had met her at a fish hut. Yes, first date at a fish hut, dropping a computer down to the bottom of the sea. That was a critical mission, the drop was a test to see if the computer leaked, saving us days of work. But more importantly, David met Diana on that critical mission. He very much wanted me to meet her, but we were busy out on the ice that season and I only saw her a few times back in McMurdo town. After David flew back to Christchurch, New Zealand, in December I found his wisdom teeth in the back of a lab refrigerator. We had wanted to leave them behind in McMurdo Sound.

So I said, "Oh, Diana, I meant to give you his teeth."

With a glint in her blue eyes she said, "I'll take care of that."

It was late in the season and the sea ice was starting to melt. I didn't think much about it, but later Diana told me she had to sneak out from McMurdo at 4 am to drop the teeth into the watery depths without MacOps noticing.

She met up with us a few weeks later. David was clearly happy to see her as they headed off backpacking in New Zealand. I didn't know if I'd see her again. After a few months of travels, she managed a transfer to Harvard, so she and David were able to sort out how to be together in the northern hemisphere, in the same city. It took another eight years for them to get married, and two Zapol-Laird grandchildren and 29 years later, I imagine his teeth are still there watching the Weddell seals and ice fish go by. Their first child, Ruth Karoline Zapol, carries the name of the first woman to set foot in Antarctica, Karoline Mikkelsen. I took Ruth with me not long ago when I was lecturing in Alaska, Japan and China. Elliot, her younger brother, is still too young to be my travel companion. He is a computer wizard, and, who knows, may someday become involved in science related to Antarctica.

One of the most exciting things to do in Antarctica is to get into a helicopter for a survey or to deliver supplies to camp. The whirring rotors reminded me of being in the Public Health Service in Vietnam. Helicopters are great because you can maneuver quickly and see so much from above. Sometimes we would go along the ice-edge to survey where the ice meets the open ocean. There, teeming with life, Weddell seals were chased by killer whales. When we landed we had to be careful not to get too close to the edge for fear of being mistaken for a seal. But helicopters are dangerous affairs, and I often cautioned my team that the best way to stay alive in Antarctica is to stay out of the choppers. That year on the ice with David and Diana, Bob had to fly to New Zealand with two helicopter crash victims who went down by the same ice tongue because of the unpredictable winds coming off of the glacier. We huddled around the oil stove when he got back, hanging on his every word of the story of keeping the pilot and the co-pilot alive and breathing until he could deliver them to medical care in New Zealand, before flying back on the next flight and landing in McMurdo only to learn that they had died. Antarctic exploration had claimed two more lives.

I loved the stillness and remoteness of the field camp. I've spent time at Palmer Station in the "Banana Belt" of Antarctica, which is above the Antarctic Circle. That means there is a sunset there, year round, and we do our work from boats as it is much warmer. I've made it down to the South Pole a few times, just for a few hours on a "Sleigh Ride," a turnaround trip. Representing the United States on the Polar Board, I have traveled to some remote parts of Antarctica and the Arctic. Someday I hope to make it to the Zapol Glacier: The United States Geological Survey (USGS) named a glacier after me on the side of Vinson Massif, within the Sentinel Range of the Ellsworth Mountains. Vinson Massif is the highest peak on the Continent. I think that is just wonderful. I love these pristine places where few people have been and we can get close to the millions of years of ice below us, and the strange creatures that live there.

Late at night, after we turned off the generator, we'd all lie in our cots at Strand Moraines, inside of our hut, listening to the sounds of seals calling to each

At the South Pole in 2011 as a Presidential appointee
to the U.S. Polar Board.

other under the ice. It's a little bit like listening to coyotes howl, but more musical, staccato bips and bahs... it makes me smile just to remember those quiet times.

Sometimes our mail would arrive by the *thump thump thump* of a helicopter out in our field camp, and we'd savor the crinkled pages and a picture from home. My time in Antarctica straddled the dawn of the internet, and satellite coverage in the Antarctic was poor at best. I always shipped my ham radio down with our lab supplies, and I'd set it up on the sea ice, planting a three-element beam antenna onto the sea ice and spend time on Sundays seeing if I could talk back home to Nikki. It was tough for the family—voices garbled in the ether, we'd be "patched" together by a local ham radio operator in the U.S. who would connect to a phone to call my family—usually in the middle of the night, waking them up so they could try to make out a few words from me on the underside of the earth while thousands of people listened in. For me it was wonderful to feel that I could make contact with the outside world and the family I loved dearly. But I was absorbed in work, surrounded by friends. I know for my family it was difficult to be reminded of the distance between

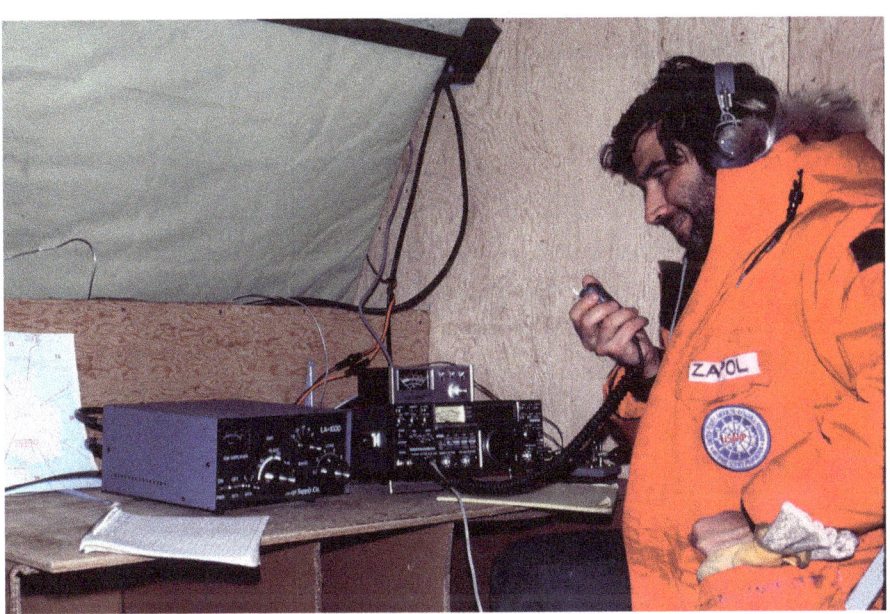

Communicating northward via ham radio.

us and often it would end with a child in tears and me sitting down to write a long letter to try to communicate what was happening and send my love and a picture back to the north.

This was the reality of doing science in Antarctica: real risk, comradeship, isolation and scientific creativity. Our science is published. I've got thousands of pages of papers out there if you want to dig in deeper to the science and measurement of fish blood pH, and the mysteries of seal breathing. But it's the seal breathing that I went to study. The brilliant evolutionary solution that they came up with is the story I am proud of our team for elucidating, and it is the story I love to tell.

After all the years on the ice, all the time away from family, and all the trial and error, all the fighting for grants and publication, all the innovations small and large, the computers we designed and the duct tape we plastered over our mistakes, we obtained a full picture of how the seals dive to such a depth as mammals.

Seals breathe the same stuff that you and I breathe. Air. What is in air? Air has oxygen, yes, about 20%. But it is mostly, 80%, nitrogen. This mix of gasses in air plays a big role in the work that I do both in treating patients with respiratory issues and creating medicine from air. These seals dive deep. Down to 1600 feet in McMurdo Sound where the Antarctic cod are 180 pounds and there are no other predators. If you are a scuba diver, you'll recognize immediately the lure of big fish, but also the risk of being an air breathing mammal at depth. If you have air in your lungs, you get the bends, which are caused by nitrogen bubbling out of the blood on ascent from a deep dive. This was originally observed in the people building the Brooklyn Bridge, and called the bends because the nitrogen bubbles cause joint pain, which causes people to bend over, holding their knees.

Seals don't have knees, but that's just me being funny—that anatomical difference wouldn't protect them from getting the bends. It is remarkable that seals don't breathe in air before they dive, because we can't imagine going down for a long breath with empty lungs. That simply wouldn't work for us. But remember, air is mostly nitrogen. When Weddell seals are diving they actually breathe out, collapse their lungs and only leave a little bit of air in their trachea. The air expands again on ascent, and that is what is exhaled immediately on resurfacing—the *pffft* we'd hear when they'd wake us up by the hole. The bubbles that trail out as they descend into the depths are a smart strategy—they are getting rid of any last bits of air, mostly nitrogen, which could cause problems in their lungs, their joints, and cause pain or death.

The pregnant mother survives not only by exhaling before diving. She also stores oxygen in her blood and muscle and, most amazingly, that spleen that serves as a reservoir of oxygenated blood, which David and I measured with Mont's ultrasound. She manages to maintain tight control of her metabolism, as Peter showed with his biochemical analyses. And all the while she keeps the baby as happy as if she were on the surface being an air-breather, maintaining rich oxygen supplies to her baby, and her own brain, heart and lungs above all other needs in the body, which was Mont's work with the rest of the team. The seal mama slowly and elegantly controls the use of oxygen so that she can spend time below ice, where she can feed in the food-rich waters, safe from killer whales and other predators at ice-edge.

I have said that David was in Antarctica with me. Both my kids were. Liza came with me on an expedition in 2006. She was an actress then. One of my favorite pictures is of my daughter Liza unexpectedly coming to meet me via helicopter in McMurdo. We had been sharing a cabin on the Khlebnikov, a Russian icebreaker. I was leading a group of Harvard alumni on a three-week expedition from New Zealand to McMurdo. Liza was taking part in on-board skits, when a passenger on board fell ill, and the ship's doctor and I evacuated the woman to McMurdo. Liza was co-pilot in the helicopter that came to retrieve me a few days later. (The woman survived and was transported to Christchurch, where she died shortly after.) Liza is a brave, excellent travel companion.

As I think back, I have one very poignant memory of returning back home. It must have been 1985, Liza was seven years old, and I had grown a big, heavy, black beard. I remember getting off the plane, and Liza ran right past me.

"Dad! Dad! Where are you?" she cried.

We got in the car, and the kids were crying, and everybody was asking what had happened to Dad? As soon as I came in the door of the house, Nikki handed me the shaving cream. I have many thousands of pictures from Antarctica, and then there are pictures of me shaving that beard off. And the picture of us all happily back home together is worth the thousand I took in Antarctica.

With Nikki in 2008 on a Harvard-sponsored Antarctic expedition.

CHAPTER 4
Exploration: Malaria

✳ ✳ ✳

I ALWAYS HAVE HAD A SINGLE mindedness about my goals, not to mention the chutzpah to reach for goals that were almost impossible. It started early and never went away. As an eight-year-old in Brooklyn, I wanted to build a rocket, as an Antarctic explorer I wanted to measure seal dives to the bottom of the ocean, and as an old man with lung cancer I wanted to ride my bicycle one hundred miles in a day. Setting a nearly impossible goal makes others take notice and join in the effort, testing their bodies and minds to take on a big challenge. On an adventure, danger is always present, mistakes are inevitable, and that's how we learn. My most formative challenge was one when I was twenty—the one which may well have turned me into a man.

I skipped my graduation from MIT in 1962 to lead the MIT Southwest Asia Expedition. The members of the expedition were a crew of intelligent kids, but not smart enough to think ourselves out of a near-impossible goal. We fancied ourselves swashbuckling adventurers on a crusade: a crusade to drive by land all the way from England to India. This trip would be nearly impossible today due to political barriers and bureaucratic red tape. Even then, this trip had only been completed once before by a group of university students, from Cambridge and Oxford. Had I done anything remotely like this before in my life? Aside from taking a big risk by blowing up my friend's backyard, no. Most of my childhood travels had been back and forth from Brooklyn to my parent's camp in the Catskills, with occasional forays to Miami Beach. I was barely out of my teens, and I had spent the bulk of the last year in a laboratory, studying an owl's eyes for my biology thesis. Now I was setting out to lead an eight-thousand-mile expedition during which I would nearly kill myself and murder someone else.

The stated purpose of this trip was to study geology, geophysics, and radio—definitely radio, one of my great passions. But my hidden purpose, and what really inspired me to dream up this voyage is that one of my friends, an Iranian, invited me to visit him in Tehran. So I concocted a plan to lead the first American college expedition to drive from England to Iran and back. I wanted a sampling of hard scientists and social scientists who would be able to contribute to this expedition. I gathered three friends to join me on the expedition. We were an engineer, a linguist, a chemist, and a biologist from MIT. Later, we added the pedigreed son of an English Lord, which seemed at the time to be qualification enough. Though he was not American, we figured his pedigree would open doors, which it did, to a point. As we asked for hosts, we advertised that we could "handle" speaking fluent Arabic, Bengali, Farsi, French, German, Russian, Turkish, and Urdu. Yes, the five of us could handle eight languages. My contribution was a shady fraction of one, but no one really needed to know that (yet). Plus, my French teacher from Stuyvesant wasn't invited.

I laid out my hand-drawn map, and charted a very long path—a careful dance above the Mediterranean and below the U.S.S.R. Then we were invited to visit the Mir of Gilgit, in Pakistan, so we decided to go even further, through Afghanistan to visit a local commander. The road maps were shoddy and indecipherable, so I got as much information as possible from the English travelers who had done it before, and I created an itinerary that encompassed 31 stops. We wanted to contribute to map building and knowledge gathering, like great explorers, like Shackleton.

Our press release (included here) helped us get some investors, wealthier classmates who gave us five hundred or a thousand dollars so we could buy the things (like radios) we needed. We convinced the Land Rover company to give us one of their cars for a thousand dollars, or some ridiculously small fee. We figured we had what we needed.

M. I. T. SOUTH-WEST ASIA EXPEDITION

Left to right: Jerome A. Winston, Robert A. Knisely, Warren M. Zapol, Norman P. Soloway

JUNE 11 - SEPTEMBER 11
1962

April, 1962

Last summer, an M.I.T. senior and a Sophomore from Clare College, Cambridge, were asked by an Iranian to "come to Teheran and visit my home." Late that night, the idea still seemed improbable, but was twice as enticing. The journey has lengthened, and the party has grown; the trip is a certainty now, but it has lost none of its original appeal.

As of early April we are torn between one Long Land-Rover and six people, or expanding to include another Rover, more people, and the advantages of mutual support. There is certainly no lack of applicants.

We plan to fly to London, pick up the Land-Rovers, then on to Paris, Rome, Istanbul, Teheran, Lahore, New Delhi, the palace of the Mir of Gilgit, and Amritsar, with our return possibly through Afganistan. We may also get the cars into Russia, depending on Intourist.

We would like to stay longer, but most of the Expedition returns in the fall to graduate schools, including two medical schools, a law school, and one postgraduate National Defense scholarship in linguistics at M.I.T. Our undergraduate majors at Harvard and M.I.T. vary from mathematics and linguistics through civil and electrical engineering to biology.

The Expedition received an invitation to visit the Mir of Gilgit through a Pakistani from Chittagong, who is our main guide and interpretor. Gilgit and its neighbor, Hunza, are seldom visited by outsiders, and almost never by foreigners. We hope to drive, but if the roads are impassable, we may have to be flown in by the Indian Air Force.

As well as visiting Gilgit, we plan to document the route carefully, providing up-to-date information on roads, rivers, and villages, and the availability of fuel and water. Extreme conditions vary from Dasht-

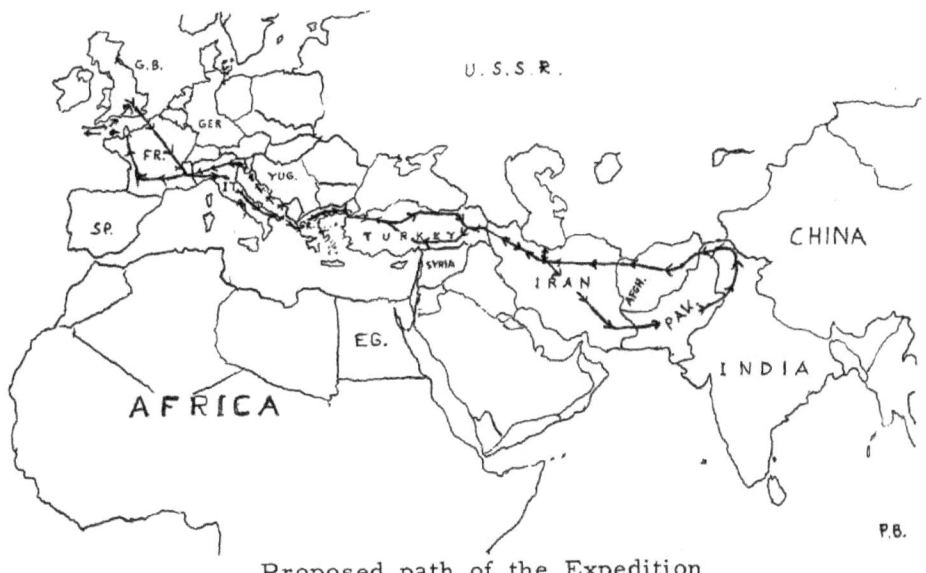

Proposed path of the Expedition

i-Lut desert at 120° F. to cold nights in the Hima-
layas of Northern India.

Information and advice are being drawn from
many sources, among them the book <u>First Overland</u>,
the record of the <u>Cambridge and Oxford Far Eastern
Expedition</u> of 1956-7, and several other voyagers who
have made similar journeys since then. We are following no one's path, but their experience with cus-
toms and terrain is proving invaluable. We are again
demonstrating the rugged versatility of the famous
Land-Rovers, and they in turn are partially supporting us, as well as giving advice and assistance with
organizations interested in the movies, tape recordings,
and photographs we will bring back.

We have experienced still and cinephotographers,
competent mechanics, good drivers, and a great deal
of linguistic talent. Among us, we can handle fluent
Arabic, Bengali, Farci, French, German, Russian,
Turkish, and Urdu. We will be the first American
College Expedition over this territory; Our Base-
Team consists of enthusiastic graduates and under-
graduates of Harvard, M.I.T. and Wellesley.

We are currently seeking assistance and advice from companies and organizations varying from camera and camping-goods manufacturers to the State Department and national and international news publications.

Mailing Address:
    M.I.T. SOUTHWEST ASIA EXPEDITION
    Box 6181, Baker House
    362 Memorial Drive
    Cambridge 39, Massachusetts

Our expedition pamphlet.

Later on I'd venture to Antarctica, to Russia, to Vietnam, Korea, Uganda, and Japan, and into the depths of the mammalian lung. For this drive, I was a considerably less old hand, and less well-equipped. It wasn't a matter of preparation—I collected all the available maps, I read the accounts that preceded us, and I connected us to contacts along our itinerary. But this was my first real adventure, and there were wide chasms in our plan. It took an arduous voyage and cost me my health to learn the lessons required for future leadership.

Though I hadn't traveled much, I felt secure about exploring the world outside of Brooklyn, beyond MIT and Cambridge. Ham radio helped me imagine that the world was in reach: I was regularly connecting to my buddy in South Africa all through college, and talking to hams in hundreds of other countries. I held postcards from people around the world, so I already felt a part of a global web, even if when I spoke to them I remained stuck in a radio shack in Bloomingburg, New York.

Sure, I was scared, but I had to confront my fears. I don't really know where I learned to take risks like this. Perhaps it was my father, who wanted the upside of life, and was willing to take a risk for it. For him, these risks were mostly financial: gambling on real estate deals, investing in watches that might get a real return or might flop. Certainly it wasn't my mother: In her mind, another Depression was just around the corner. She worried that we would be penniless and things wouldn't go well. She feared a life of poverty. I'd never been to Europe, though my grandparents had all been born there. Overall, I had led a pretty sheltered, middle-class life in a Brooklyn Jewish enclave. I wanted to do more with my life.

Everyone from home thought I was crazy, including my parents. During my last regular Sunday call home before I left, I could hear my mother's worry:

"So you're going."

"Yes mom, I'm leaving tomorrow."

"How are you getting to Europe? Swimming? I'll never hear from you again."

"No mom, I sent you my itinerary. We found cheap tickets. I'm traveling to Ireland, and then Iceland, and then London. To stay with Joe Silk."

"That nice English boy who was a counselor with you when the campers tried to murder you with knives?"

"They were throwing knives mom, not at me. Yes, that Joe. We're staying in his dorm room."

"Are you going to Germany?"

"Yes mom, I told you, We're stopping in Germany. I love you. I'll write. I promise."

"He says he'll write. No he won't. He'll get murdered by the Germans or he'll fall off a cliff in Switzerland or marry a *shiksa* in France. Death! I'm putting your father on the phone. Ben, come here, it's your smartypants *meshugenah* son."

"Warren, we love you. I sent you some money."

"Thanks dad." My voice wavered. "Look, I should go."

"Bring back stories," he said. I could hear the pride in my father's voice.

And so the first four explorers that made up the MIT Expedition, young but brave (or at least brave enough) set off. Once we got to England, I was delighted to see Joe Silk, and we crammed into his Cambridge dorm room. Joe would become a famous astrophysicist of the early universe and an expert on Big Bang Theory, and he always had a wonderful, otherworldly intelligence. He would also soon introduce me to my wife Nikki, a key to my universe.

Then we met up with our fifth explorer, Oliver Walston, in the grand London apartment of Lord Walston, his father. I had English teatime for the first time ever, brought to us by a butler. I enjoyed the Lord, who had an inflated sense of self and entertained us with stories of his own travels. But I was distracted by the butler and the other staff, overly-mannered waiters with arched eyebrows who seemed to be amused by my bad American manners. I had never met anybody with a staff, with servants. My version of *Upstairs, Downstairs* was when my *bubbe* and I would bring my grandfather schnapps when he lived upstairs. This was entirely different. I wasn't scared to come off as uncultured. I had no choice! I was! And I wasn't embarrassed by their arched eyebrows—there was no offense here that a good self-deprecating joke couldn't disarm. I gained a sense of the life of privilege and entitlement that Oliver Walston came from, and at that point I was still enjoying the association. Even though this trip to London was already a journey to an unknown place, I looked forward to the road getting a bit rougher.

We bid adieu to our manservant and made our way to the Land Rover plant outside Birmingham, England, where we were trained by Land Rover to drive and service their car. My role in the team was as the guy who figured out the car and where we were going. Because I was the main driver, I was the most attentive at the Land Rover headquarters. We then flew to Belgium to meet our trusty car on the European continent.

In the picture of us in the brochure for the expedition, we are posed around the Land Rover, dressed in white dress shirts and black jackets, and black square framed glasses. We looked the way you might imagine MIT students in the early '60s looked. We all appear upright, uptight, hopeful, with a competitive gleam in our eyes. As expedition leader, I sat on the hood of the car while the other guys leaned on it, jutting my shoulder out, emulating James Dean. In reality, I look like James Dean's pre-teen little brother! I always looked younger than I was, and I <u>was</u> a young *pisher*, who began at MIT at 16.

We drove to Paris and the French countryside. In a letter home, I indicated that the aroma in the Land Rover might have quickly turned a little foul: "Parisians don't believe in hot water so it is difficult to keep clean.... The people drive horribly too." However, I definitely enjoyed the delicious wine and cheese, which made for a great budget meal and was my introduction to the pleasures of non-Kosher wine.

We then crossed into Switzerland, stopping briefly in Germany. My mother's fears echoed in my head, and I thought of my father who studied medicine there until he was kicked out for being a Jew. And I also shivered as I remembered the tattooed numbers on the arms of Holocaust survivors. I looked carefully at the world around me for traces of the horrors that had happened. I could see that many buildings had been completely bombed and remained destroyed, while others had been completely rebuilt. We were headed to a small town in the Black Forest, where there was a mayor who said he wanted to meet us. I was afraid. Should I hide that I was Jewish? Was that even possible, with my nose and furry eyebrows? Could I hide behind my *goyish* travel mates, perhaps Oliver Walston?

When we drove into the town center, I was amazed. The town center was decorated with ribbons and lights, there was a banner for us, the villagers were gathered to greet us, and the mayor hosted a press conference. The local press were eager to hear about our adventures thus far and where we were headed, and the townspeople gave us good beer. Everyone was so nice. Could this be a place where horrors occurred?

Later in the night, the portly mayor leaned in to me, his breath heavy with sausages and beer. "Where are your people from?"

"Uh me?" I gulped. "Brooklyn."

"Hahaha. *Nein*, before zat?"

"Austria and Germany. Old Russia…" there was a long pause. I knew what he was really asking, of course. I looked around, as my pulse raced. There were lots of people around, including my buddies, though we were outnumbered. I charted an escape route to the Land Rover, and figured I could make it in time. "I'm Jewish," I finally said, and braced myself for the mayor's response.

"Ah! Yes, you are Jewish!" He looked hard into my eyes. I grasped the car key in my pocket, ready to bolt. His face broke into a smile. "That is good. Now I have a Jewish friend." He offered his hand to me. "We need more Jewish friends."

I shook his hand, amazed. Could Germans change their antisemitism that quickly? It had only been 18 years since the end of the war.

"*Gute Fahrt!*" My new German friend shouted down at me. I looked at him, hugely embarrassed and perplexed. Had I farted, in my terror? Had I made a different kind of gas chamber? He must have seen my red face and roared with laughter. "Bon Voyage in German! *Gute Fahrt!*"

I told my travel mates this story later that night and they all roared with laughter. Though they weren't Jewish, they all grew up with the portrayal of the "ruthless krauts" and were similarly relieved to find a welcoming village in Bavaria. And a fart joke can always dispel tension.

As we drove East, past the breathtaking countryside of Bavaria and into the Alps along Hannibal's route, we thought of traveling through Yugoslavia. Maybe life under the Iron Curtain wasn't as scary as I thought, and we would be greeted as heroes there too? A conversation with an older English gentleman at a Hofbrauhaus dissuaded me. A ham radio nerd like me, he warned me that the Yugoslavian border police had confiscated his friends' radios. I didn't want anything to happen to my precious radios! So instead we continued down to Italy, and took a ferry across to Greece.

In Greece, I had my first hot shower of the voyage, and it was glorious for me and for my carmates, since my odor was drawing attention. I began to run out of money, and, to my dismay, one of my wealthier crew mates spontaneously abandoned us to go visit Athens, while the rest of us soldiered ahead for a few hundred miles, with more silence as we rode. This soured me as I looked out along the dry Grecian landscape. Why was I losing team members? Why were people so willing to act in their own self-interest, knowing they were making it harder for others?

I began to write more letters home, as promised, sharing reports about the cleaner bathrooms and a description of the dark Turkish coffee with coffee grounds that would get stuck in your teeth, that reminded me of the thick coffee that my Grandmother drank. I knew they would relate to these details, though of course there was so much more I was seeing and thinking which I didn't know how to describe to them, like my comical meeting with the German mayor, and the attrition of my team. Brooklyn felt so far away. I became homesick, though I would never admit it to my travel mates.

Our impulsive traveler rejoined us in Istanbul, and breaking the silence, some grumbling among the group started. Oliver Walston was argumentative and difficult, used to better conditions. He didn't want to help maintain our Land Rover with our rotating chores, and his temper would flare up when I insisted he help. He treated me worse than his servants, and then my temper would ignite too. Some of the crew I loved, some spurred me to healthy competition, and some plunged me into anger and envy. To add fuel to the fire, I was somewhat hungry—since my money was dwindling, I'd try to space out my major meals to once a day. I'll confess, my anger ran hot when I was hungry. The long drives would become unpleasant.

From there we drove into Iran. At this time in the early 1960s, Iran was a progressive, secular monarchy, and the Shah had close ties to America. However, as we arrived in Tabriz, a city in Iran's Northwest nestled below pointy volcanic mountains, local goodwill started to erode... to say the least. When we stopped for gas, the attendant leered at us and stared at our Land Rover.

"Sour puss!" One of my team muttered under his breath, and we tried to hide our laughs.

Then, suddenly, a rock hit the side of our car. We looked around, and couldn't find the culprit. But old men also glared at us from across the street, and we heard yelling from down the road. To our horror, a crowd of young people with sticks was gathering and appeared to rush in our direction. "What the hell?" I exclaimed.

"They're after us!" Walston yelled. "You've led us into the enemy's mouth, you idiot!"

I snapped to attention as a stream of expletives spilled from my mouth scattershot—at Walston, at the yelling crowd, at myself. I goosed the engine and the attendant jumped away, and we peeled out of the gas station. As we drove away the crowd followed us, but we outpaced them. We were dumbfounded at first, then we realized what had happened. Our British Land Rover had incited a demonstration. Iranians had a deep mistrust of the British, who had been ruling Iran in one way or another since the mid-1500s and plundering their resources. "Bloody Brits!" I muttered to myself. Imperious, Walston shot me a look. He was the enemy, not us. And so was the Land Rover. How could we explain that to the growing riot? Thankfully we didn't need to, as the crowds were receding in the rearview mirror.

In Tehran, I found some relief from the conflict in and out of the car. I took a few days to myself while the rest of the group stayed at the British Embassy. I visited my friend Nader—finally fulfilling the invitation that was the inspiration for the whole trip. He was studying to be an architect, and his family had a beautiful home. I was overjoyed, so happy to see a familiar face, and to meet his kind family. It felt, just a little bit, like I was at home with them. Their unconditional goodwill was a brief oasis from the testy relations I had been experiencing with my travel mates, and the wariness required when traveling in unknown lands. And I feasted on delicious foods I had never had before, like aloo baloo palo, a chicken and rice dish augmented with the unique ingredient of sour cherries. I hadn't had a real meal in weeks. I wished I could eat that dish forever, and I still do.

When I returned to meet up with my travel mates, I discovered that Walston had left our trip and returned to London. I was shocked. I supposed the small demonstration had shaken him, and his trust in my leadership, and I think he felt vulnerable as the sole Brit in the group.

I'll admit, it was a relief, but I feared losing his access. With his father's connections, he opened the door to embassies across our travels. I was nearly penniless: I had about $100 left to my name. I wrote home: "Please send money to Tehran or I will come back hitchhiking." I'm sure my anxious, doting mother was thrilled. She sent money via Western Union to Tehran which would later become key.

At the Caspian Sea, we spent a day cooling off from the 105 degree weather by swimming in the brackish sea. That evening, in our tents, everything took a turn for the worse. I was bitten by massive, nasty flies, and I could hear my friends shaking around in the tents, fighting off the flies too. I barely slept, scratching, slapping, and cursing all night. The next morning, as I was driving 60 miles north of Tehran on the way to Qom, I developed a fever. Imagine: You arrive at Qom, and while trying to appreciate its spectacular mosque, you're shaking with fever and chills. My fellow explorers did not seem alarmed. None of us had any medical knowledge. I drank endless cups of tea and shivered through the night. When I got hot again, I slipped into a cool bathtub. At the time I figured the fever would pass—it was just some of the roadside food I ate.

As an indestructible young man, it did not occur to me that I, like everyone else, was fragile, and that I had contracted a life-threatening disease. The fever kept coming and going, and I was losing weight fast—I just couldn't seem to shake it—and exhaustion became a constant state. I was sluggish and sleepy, and always thirsty. We carried ten gallons of water with us and used it up daily. I had to give up driving duties—I wasn't strong enough, and my increasingly bony body bounced around in the back of the Land Rover, barely registering when we had entered Afghanistan. We visited Herat, Kandahar, and Kabul, then took the famous Khyber Pass into Pakistan. In Rawalpindi, Pakistan we camped on British Embassy Grounds. Sleeping on the hard earth, sweating and shivering through the night, I became delirious. I passed out.

When I regained consciousness, I saw a woman gazing down at me, with a beautiful halo. Where was I?

I wondered out loud: "Am I dead?"

The woman smiled.

I smiled back. She was beatific. Did her smile mean yes? I assumed so.

"Are you an angel?"

She laughed. I enjoyed the sound of her laughter.

Then she spoke, and she explained everything. I was alive, and she was a doctor at the Holy Family Hospital and had saved my life. Her halo was her nun's habit. She was a Harvard-educated American religious sister. She explained that she had sampled my blood and spotted the cause for my illness under a microscope: the parasite that causes malaria, as well as amoebiasis, a separate infection.

She quickly prescribed a treatment of chloroquine phosphate, and perhaps something for the amoebiasis. To my relief, my fever broke and my health turned around, fairly rapidly. I regained energy and appetite. Without her sure diagnosis and treatment, the condition would have continued wasting me away to death. I had lost 35 pounds already and had tightened my belt to a tiny ring. I wasn't mad about the weight loss, but I would never recommend a parasite diet. This angel doctor, whose name I sadly forgot in the delirium long ago, cured me. I owed her my life.

When I was released from the hospital, I spent time in Pakistan recovering, and my MIT friends helped me. We camped at the Atomic Energy

Commission, of all places. Pakistan was building a nuclear bomb at the time, and someone from MIT knew some people there. At this point, I had no stamina or funds for the long drive back to return the car to England. In fact, I didn't really have enough money to return home at all, so I planned to sell the Land Rover. I made a deal with a rich farmer in a nearby city who wanted to buy the car, and I felt like I had a solid plan to get home within a few weeks.

I was bumping along at 30 miles an hour on a muddy road between the farm and Lahore, dreaming about getting home to Brooklyn and eating a real BLT. After all this time in a Muslim country, I really missed bacon. As we passed under a bridge, I suddenly noticed a movement out of the corner of my eye. From behind a column, a figure darted out in front of my car: a young boy, about eight years old, dressed in rags. I tried desperately to break, but unable to stop in time, I hit him with the front left of my car, and then stopped. Voices echoed in my head: "Don't stop if you get into an accident!" Our friends in Rawalpindi had warned. "The locals will rob you, they'll kill you." I was terrified! Had I hurt or killed the boy? I quickly got out of the car, and found the child lying in the road, his head and leg bleeding. I looked around—no parents, no friends. What was I to do? I picked the little boy up and carried him into the truck where I laid him across the back seat and drove down the road, looking for help.

I finally flagged down a local who guided me to a nearby hospital. I carried the crying boy in, and the doctor there spoke to the boy in Urdu, and began to clean and treat his wounds. I spent a nervous, helpless night at the hospital, watching anxiously as the doctor bandaged the boy's scalp lacerations and put his leg in a cast. I was in a state of anxious paralysis, a feeling I have avoided ever since. I was unable to help someone who was hurt, and even worse, hurt by my doing.

A policeman biked over to the hospital. He must have been alerted by the doctors, or the local who guided me to the hospital. He spoke with the boy in Urdu, and then turned to me.

"Passport?"

I brought my passport out of my bag, my hands trembling. He eyed me, and took my passport, and walked away. As I saw him mount his bicycle and ride off, I was terrified. Would I be detained or jailed in Pakistan, emaciated, exhausted, and nearly a murderer? Would I ever get my passport back?

He returned after a short eternity.

He said I wouldn't be prosecuted. Hallelujah! However, I needed to pay twenty dollars directly to the family.

I didn't even have twenty dollars to my name, but I didn't tell him. I promised to give money to the family within the month.

I was so guilt ridden. The police didn't blame me for the accident, strangely. "Allah willed it," they said, but this further confused me. Why would anyone believe Allah willed this? Couldn't this have been prevented by humans? By building sidewalks, and road barriers, and keeping children out of the street? But I didn't argue. I was the outsider, and I just wanted to get home.

How was I going to get the twenty dollars? At the Atomic Commission, I called my old friend Nader in Tehran. "Did you get a letter from my mom, Florence Zapol, for me?"

"Yes, Warren, I've been expecting your call. How are you?"

"Awful, Nader, I've never been worse. But please, could you get that letter and open it right now and tell me what's inside." My heart was in my mouth. I listened as he ripped open the letter.

"There is a letter, and money."

"How much?"

"Let's see. Five hundred dollars."

I sighed with relief. We hatched a plan for me to get the money, and after an anxious week of daily trips to the bank, it was in my hands.

I delivered the money to the boy I had hurt, and his mother. Thankfully the boy was recovering quickly and would soon be walking normally. Possibly because "Allah willed it," but more likely because he had proper treatment from a capable doctor. I was relieved. I visited the local police station and I got my passport back.

Finally the paperwork was complete to sell the Land Rover, and with a bit more money, I went shopping with the other MIT travelers, picking out the perfect gift for my mother. I found her a sari to wear in Miami Beach (which she would never, ever wear). Then, I prepared to return to America, parting ways with the remaining MIT travelers. The trip had been hard and harrowing, and yet even near death I was so grateful for the adventure and the lessons. In my last letter home, from Abbottabad, I wrote: "Adventure is hardship in retrospect."

During my flight home, I reflected on the photograph of my group around the Land Rover, and our reasons for setting out on this voyage. We were all individuals, with a common goal: to widen our own understanding of the world. On those terms, I had succeeded, and I know others did too. I had new friends who spoke languages I had never heard of, people who I had assumed were German villains, but who turned out to be kind. I learned about worldviews that fundamentally challenged mine: The words "Allah willed it" bungled my rational mind.

But the rosy picture of the Land Rover crew clouded in other ways. As a group, we weren't united, especially when hardship hit. Walston fled, others avoided responsibility, no one really saw the warning signs of danger in Iran, or with my health. We didn't create the extensive documentation of our route which we said we would: "Providing up-to-date information on roads, rivers, and villages, and the availability of food and water." In retrospect, the only lasting documentation I have are the letters home to my parents. As a leader, I was disappointed in myself. I knew the next trip I led would need to be more clearly oriented around a common goal. Not a selfish goal, but a united goal, oriented in meaningful research.

When you take an adventure, you can never eliminate risk. But you can make sure you have a great team of experts and a strong support team, with everyone carrying their weight. I learned that on this trip, and on all remote expeditions after it, each team member will likely be pushed beyond their comfort, and have to participate in menial tasks beyond their area of expertise or interest. I didn't have any patience for those who don't do their share of work, the Oliver Walstons of the world.

Research is a risk. What you're doing might be worthless. What you're doing may cost lives. You don't really know when you start out. But since this expedition, I have been much more careful in laying out the path, the road ahead, in hopes that by the end, I might be able to contribute to humankind.

I couldn't stop thinking about the doctor I had mistaken as an angel, who had saved my life, and about the helplessness I felt as I watched the boy suffer in the hospital. The doctors took action swiftly, decisively—knowing what to do, and how to manage the risk. In my treatment, the doctor's ability seemed logical, all because she knew exactly the common disease she was dealing with. She was able to act when my life was hanging in the balance. I was in awe of her brilliant diagnostic and therapeutic skills. I wanted to know what she knew.

CHAPTER 5

Death Defiance

* * *

CONVALESCENCE IS A HELL OF a thing. A person might appear to recover from an illness, may in fact no longer show any symptoms, and yet traces of that illness may remain. In a culture of medicine based on a symptomatic approach, these sequelae are often overlooked, often written out of the narrative. We've gotten in the habit of taking medicines when something is painful. When the pain subsides, we say we're better. But compared to modern medicine's efficacy in taking immediate therapeutic action, it is weak in dealing with convalescence, physiotherapy, and long-lasting consequences. A person's body and mind need time, peace and care to fully recover from any illness and to adapt to any long-term effects. Malaria not only left its mark on my body—never again would I be allowed to donate blood—but it also had a profound effect on my life's work.

On my flight back home from Pakistan, I thought about what I was bringing back with me. I had a trunk full of souvenirs and tchotchkes. I had cheap clothes and half-written postcards. I had some beautiful stamps for my collection. Most importantly, I had a newfound clarity regarding my own future. I knew where I was headed as sure as I knew where my flight was headed: I wanted to apply for medical school. My life had been turned around, saved, by that woman gazing down at me, the woman with a beautiful halo who tested my blood, saw parasites, and saved me. I had not been a stellar student during my time at MIT. I'd spent most of my time on the ham radio. I'd taken "physics for poets" so I could learn the big picture and earn a passing grade. But now I knew I had to try. I had been saved, and if I could save even one person in the same fashion, it would be a debt repaid.

Back from Pakistan, in my parents' home in Miami Beach when
I told them of my decision to pursue medicine, 1962.

It was August when I returned. The regular admissions cycle for medical schools had passed. I had to do something unconventional otherwise I'd end up back in my parent's house, and I couldn't imagine anything less exciting after the life I had been leading. I decided to just show up at the Boston University Medical School admissions office. That was my first step and, I hoped, my last. When I walked into the admissions office all the heads in the room turned. I'd lost a lot of weight in Pakistan, and not put all of it back on. My pants still barely stayed up. I could tell the people in the admissions office were surprised at how gaunt I was.

Smiling weakly, I asked to speak with the admissions officer, and was ushered into an office with a bespectacled gentleman who peered out from behind a giant stack of paper.

"How can I help you?" he asked, trying not to stare at me.

I explained who I was—that I had almost died in Pakistan and a Boston-trained doctor had saved my life. "So, I want to be a doctor."

I handed him another copy of my application.

He leaned back in his squeaky metal chair and flipped through the papers. Finally, he spoke. "Well, we don't hear that everyday…"

I wondered if he was going to throw me out. Was he persuaded by my conviction? Had he heard this kind of thing a million times before? Did he think my personal story was irrelevant or even inappropriate? Maybe he thought I was nuts, suffering from cerebral malaria, and was about to call security to remove me.

He flipped over my application and smiled.

"Why don't you take the MCAT and call us with your scores as soon as you get them?"

That was the best outcome I could imagine. I've always had a knack for testing. I crammed as best I could, took the test a few weeks later, and when I got my scores they were good enough to embolden me further. I called the bespectacled admissions officer.

"I did pretty well," I said.

"Let me see what I can do," he said.

They admitted me in November. I jumped into the classes late but grateful.

Medical school changed my life. Maybe that's too obvious to even say, but it illuminated a bright path for me. My first dissection in anatomy class is etched indelibly in my memory, not only because it was my first view of death up close, but because it came so close on the heels of my own brush with severe illness.

Anatomy is fundamental to understanding medicine, and a first-year dissection is a baptism by fire. There you are, faced with a cadaver—a body that belonged to a person who once was every bit as alive as you, and who elected to donate all of themselves to the advancement of medicine. That's humbling. People who do this make an incredible contribution to education and science. Am I worthy of this person's afterlife?

Cadaver dissection is also a necessary filter for aspiring doctors. Those who can't stomach dissections drop out of medical school completely. They should. That might sound a little callous on my part, but remember that it is the job of the doctor to look unflinchingly at the human body, at all its working parts and all its flaws. There's no place for squeamishness when it comes to studying the dead, because they hold the keys to life.

The bright room smelled overwhelmingly of formalin, which is noxious stuff and irritates the throat. I stood with a fellow freshman med-student, both of us fully gowned and gloved in front of our covered cadaver, one of some thirty in the room. We'd heard plenty of stories from the second-year students about how this would go, and that was how it went—we each uncovered an arm to begin dissection, and gradually I worked my way with my partner around the body of an elderly lady. You never forget your first cadaver. Of course, we were never told her name, never learned anything about her life, were not even told how she had died. But her body told us tales about her physical condition and history of disease, and by extension about her life in general. I looked to see which bits were missing. Did she have an appendix? Yes, she did. How about her uterus—had she had that removed? Nope. This lady was in good shape (except for being dead). I enjoyed feeling like Sherlock Holmes. I was truly fascinated, and I had so much to learn.

I thought of my father Ben often during those early days of medical school. He too had started down this road, only to be blocked before he could complete his studies—stopped by the workings of history itself. As I studied Johannes Sobotta's famous anatomy textbook in medical school, poring over its fantastically vivid and delicate drawings, I thought about the events that had brought me there, and the much larger events that had kept my father from doing the same. He had been the victim of bad luck, but the beneficiary of good luck, too. His dreams had not survived the Holocaust, but he had.

I remember calling my father one evening.

He was kvelling, "I'm so proud of you, Warren, for studying anatomy. Tell me, have you gotten to the foot yet?"

"Thanks, Dad," I said. "Not yet, but I thought the arms were remarkably complex."

"I just wish I could be in class with you," he said wistfully, "But one tuition is plenty!"

At Boston University, once a week we had a lecture from an unusual figure in a tuxedo and bow tie. He was a celebrity, not only on campus but around the planet: My chemistry class was taught by none other than the world-famous science-fiction author, Isaac Asimov. Asimov, whose works helped to create the modern science-fiction genre and who personified modern science-fiction, had taught at Boston University since the late 1940s. He was dazzling. It was not surprising that he had a gift for communication, but he was a remarkably good teacher who made an indelible impression on us. I have to admit that I was pretty star-struck. Being in the presence of this visionary author brought me right back to my childhood days of heady imagination, setting off rockets in Brooklyn fields and establishing radio contacts with operators and strangers halfway around the world. Hanging on every word of every lecture of Asimov's, I remember feeling, no, *knowing* that I was on the right trajectory through the universe and I was determined to make the most of it.

Other than anatomy and Asimov, the classes at Boston University were serviceable. I was grateful to be in medical school—any medical school—but much of what I was being taught didn't interest me much. The material seemed dry, the professors even more so. And then there was the cost. Though my father was always supportive, he didn't hide the fact that paying for a fifth year of education was taking a toll on him and my mother. Determined to continue my studies, I applied for a New York State scholarship, and to my astonishment I was selected and was admitted to the University of Rochester. I know this made my folks proud and I was relieved to lift the burden of my tuition.

While I was studying medicine at Rochester, I was also able to return to my passion for electronics. Electronics was a rapidly expanding domain at Rochester, the birthplace of Xerox and Kodak. I was able to add electrical engineering elective courses to my medical studies, eventually making it most of the way to a masters degree in the subject. Electrical engineering would be a critical element of my teaching and research for the rest of my life. I often have broken down physiological problems into the language of electrical engineering. The flow of blood in the lungs and the flow of electrons in a circuit share many similarities. It's beautiful how a deep understanding of basic principles of resistance and flow can be essential across disciplines.

One day in the fall of 1963, my pathology professor at Rochester, Joe Martin—he would later go on to become the Dean of Harvard Medical School—stopped into our classroom and spoke in a solemn but shaky voice. "Ladies and Gentlemen," he said, "I have the sad duty of informing you that the President has been shot."

The normally jocular, warm atmosphere in the room was shattered as if by a bomb. Like Hiroshima before it, like the lunar landing or 9/11 or the start of the COVID-19 pandemic after it, nobody who was alive at that time forgets where they were when they found out. In the immediate aftermath of Kennedy's assassination, we had no idea who was responsible, or what the immediate repercussions of this event might be. Conspiracy theories swirled. Could it be a Soviet plot? A prelude to a first strike? This was one of the "hottest" moments in the Cold War. A nuclear first-strike by the Soviet Union was a very real worry in people's minds. The USA and the USSR were creating ever-more fearsome weapons of mass destruction. In those days, when the Soviet Union was trying to deploy them in Cuba, in our backyard, and missile launch-capable submarines on both sides were prowling the world's oceans, it felt as though we were edging ever closer to an incident that would trigger global nuclear war and the end of civilization as we knew it.

In this climate of existential turmoil, the only way to survive and flourish was to focus on the everyday details of life. This was different for each person. For a parent, it might have meant focusing on the happiness of children. For a husband, it might have meant focusing on the happiness of a wife. For me, it meant figuring out the future course of my education, which meant exploring a number of fields in order to choose my medical specialty. Rochester's medical school was hyper-focused on psychiatry, an approach to medicine that did not appeal to me. I was interested in harnessing my strengths in electronics, physics and physiology. But I was also sensitive to which courses excited me and which left me cold. My favorite class was pharmacology, taught by Professor Harold Hodge, who wrote our textbook and would tell daring stories about working with the FDA to fight off bad medicines: he was instrumental in keeping thalidomide out of the U.S. and banning cigarettes. I also enjoyed a clinical surgery rotation taught by Sy Schwartz, who encouraged me to pursue surgery. In general, I was drawn to medical problems that I felt I could fix.

At Rochester, I also made some friends who would have a profound impact on my thinking. These were the medical school classmates with solid science backgrounds: Wayne Myers, Donald Parsons, Larry Aronson and Michael Geoffrey Rosenfeld. The physician-scientist mentality we shared helped to guide me through school, and I realized that not everyone was like us, even among medical students.

During summers, I branched out, explored more options. I spent one summer in the Boston University gynecology department working with Martha Frigoletto, building an electromagnetic flowmeter for the fetus to measure the flow of blood. Such precise measurements would become critical in my later scientific endeavors and clinical work, where we benefited enormously from new monitors like the pulse-oximeter (which would be developed by colleagues of mine in the coming years). Unfortunately, this summer project was to remain an uncompleted project, neither the first nor the last of my abandoned projects. At the time it was simply beyond my skills to finish it.

There was another pediatric incident in that period that shook my world, though I didn't realize it at the time. A few months before President Kennedy's own untimely death, his fourth child, Patrick Bouvier Kennedy, was born prematurely with severe hypoxia on Cape Cod, and transported by helicopter from Hyannis to the Fenway Gardens and then by ambulance to Boston Children's Hospital. A blue baby, he died on August 7, 1963. This day would play over and over again in my mind during the many years that I devoted to solving the challenges of treating blue babies. The emotional cataclysm of losing a baby reverberated in my mind—and over time that feeling never diminished. This feeling drove me to ask: Why would a baby not turn pink? What was not working in the lungs? Later, parents would approach me and tell me that their son or daughter had been saved by my invention, and it would always bring me back to August 7, 1963, and the question of whether nitric oxide could have saved Patrick Kennedy.

In my fourth year of medical school I had to apply and interview for internships to advance to the next stage. This was a tense moment in my medical studies. Inspired by the surgeons who I had observed, and persuaded by their advice, I wanted to go for a surgery specialization, but I had major anxiety. Was I good enough to do it like them, in a great hospital, with all the pressure of highly complex cases and demanding colleagues?

I don't know whether this anxiety contributed to my performance in the exams and interviews, but I remember those hurdles as one of the toughest challenges I had ever faced up to that point. When I applied for the MGH internship and residency on surgery, my interviews with members of the committee didn't go well. My first discussion was with Paul Russell, chief of surgery, who was a leader at MGH of such stature that his portrait hangs in the hallway to this day. (Some of the work I have done across the decades has been exhibited in that hospital's museum, which is named after Paul Russell). Russell was an intimidating figure to a young medical student, and he looked at me with his steely eyes. "What is the longest living cell in the bloodstream?" He asked me.

I remember pausing for a second at least—not long enough, as it turned out—and then stammering "n-n-neutrophil." I knew immediately that my answer had been wrong. My mouth felt like sandpaper and the room did a little spin. It was all I could do to catch my breath. In the pause in the conversation I remembered the correct answer was the lymphocyte. I recall feeling that the question was tough, but fair. A surgeon needs to know his or her medical stuff, not just to have a loose sense of things but an absolute command of them. Most of the time, you only get one chance to get it right, and it's often a matter of life and death. The stakes couldn't be any higher.

The next question came from Dr. Henry Beecher, Anesthetist-in-Chief: "In what position would you place a patient for post-op transport?"

No sooner had I said "prone" than I realized that I didn't know the right answer. Of course, Beecher explained, you must always place an extubated patient in the lateral position for post-op transport so they don't aspirate.

So much for an internship at MGH.

My next interview, with Dr. William McDermott, director of the Fifth (Harvard) Surgical Service at Boston City Hospital (BCH), was a totally different story. The discussion went wonderfully. What sealed the deal was my interest in biomedical engineering, which he took very seriously. Not only was I thankful to be accepted as a Harvard surgical intern, but I felt validated for my choice of engineering and medicine, which had a positive and substantive influence on my career from that moment on.

My time at BCH spanned from July of 1966 to June of 1967, and it was a hell of a year: challenging, exhilarating, exhausting. Most of our patients were Irish and African Americans—a reflection of the cross-section of Boston we were serving at that time. The medical problems they faced spanned the entire spectrum. In retrospect, this was the year that made me a doctor. What's more, it really made me believe that this was my calling, that I could do this job, and that in fact I was made for it. It wasn't easy—it took all I had, and I spent all my time sleeping when I wasn't working—but it was worth it.

Learning how to circumcise a baby was one of my more memorable lessons. In the 1960s, circumcision was becoming more and more common, and not just for religious reasons. By 1972, about 85 percent of male babies born in the United States were circumcised. I'll never forget my senior resident leaning over to me with a piece of advice. "Cut straight," he said, "or else the patient will pee sideways." I looked at him in panic, suddenly aware that my next move had long-term consequences for this baby. I had the image of this poor kid—at six, sixteen, sixty—peeing all over the toilet each time he went to the bathroom, wondering who was responsible. Perhaps the fellow next to him would have a similarly misguided stream and they would, after comparing notes ("What hospital were you born at?" "Me, too!") track me down as the shared cause of their poor aim. I later found out that one of my surgical teachers, Moses Judah Folkman—a great teacher, and a great lab researcher—was the son of a rabbi. My senior resident told me that Folkman also had trouble getting the circumcision straight the first time. That made me feel better, but only a bit. After all, rabbis don't circumcise either, that is a job for a mohel.

Judah himself had been chief surgical resident at MGH, the first Jew to hold that position, and was recruited to BCH. I was recruited to BCH by him and by two other doctors who went on to have a huge influence on my life and career trajectory. The first was Dr. McDermott, the man who had admitted me to Harvard surgery at BCH in the first place, and the other was Dr. Jack Norman. Jack was a cardiac surgeon, one of the rare African American surgeons in the U.S. in those days. He created the first abdominal left ventricular assist device, a massive contribution to the medical technology of today.

I always respected Dr. Norman just that bit more for how hard I knew he must have worked to get where he was. Not only because heart surgery is remarkably challenging and difficult, or because he was an inventor whose legacy touched many patients every day, but because in those days there was so much prejudice. Dr. Norman had to work his way to the top through a system of thought that didn't think he was good enough to be where he belonged, and what was even worse, didn't want him to believe it, either. I feel that this disturbing aspect of our society has improved considerably over the years, although there is still far to go. For my part, I have mentored and supported African American doctors as they have risen in the system with less resistance than what was common in the mid-60s. Dr. Norman didn't have the benefit of that kind of treatment. And yet he persevered and believed that the Harvard system was a great place for everyone to train and run a lab and practice medicine. Of course, there was anti-Semitism in the academic and professional worlds as well. It had been a huge factor in my father's professional trajectory, and as a result had reverberated in my childhood. But since I came of age in a time and place where anti-Semitism was not publicly acceptable, I did not feel it as acutely. Certainly I faced nothing like the "sociologic, economical, environmental and political factors that continue to dehumanize and demoralize" the Black medical community that Dr. Norman pointed to in "Medicine in the Ghetto," an early consideration of racism in the profession that was published in *The New England Journal of Medicine* in 1969.

Drs. Folkman and Norman, both brilliantly recruited by Dr. McDermott, modeled careers that combined research and surgery. The idea of somehow wedding a healing profession to the noble pursuit of science lit my fire. Without their advice and encouragement, I would never have had the belief in myself to pursue the avenues I did. I have taken this as one of the central messages that I transmit to promising young physicians and new mentors. Encouragement and guidance at just the right moment can go a long way in creating a great outcome for someone who might not believe in themselves. Self-doubt in a young physician's mind can be reinforced by negative messages from other senior physicians. I was fortunate to have these three positive role models and mentors during this critical time in my development as a physician-scientist.

But again, it wasn't all kind words and encouragement though that shaped me. It was the trial by fire that was the Boston City Hospital (BCH). For the most part, I had no reason—or for that matter no desire—to assist in the emergency room, which was a hotbed of chaos, but I was pulled into it one day on my neurosurgery rotation. Paged by the hospital, I learned that a Brinks security guard had just arrived by ambulance. I came into the emergency room to see blood everywhere, a real horror show. The guard had been protecting a bank delivery when he was shot by a thief, and my mind immediately went to the Boston Brinks heist of 1950, which was the largest robbery in U.S. history at the time. Had this Brinks guard been a hero, accomplice or a bystander? All I knew was that he had suffered an intracranial carotid tear. In other words, the main artery in his head was torn and he was bleeding at an alarming rate. My job as the young intern was to transfer him through the subterranean tunnels connecting two wings of the hospital. I wheeled him out of the ER and down into the tunnels, and—kicking as I walked to chase the basement rats away from the stretcher—I stopped in front of the elevator and hit the button for the neurosurgical operating room on the 10th floor. As soon as I hit the button, I saw the cables shuddering behind the elevator grate, but not moving. At that point I remembered Vinnie. BCH Neurology Department had a policy of employing neurological patients who were otherwise incapable of holding down a job. Vinnie (not his real name) was one of those patients operating the elevator during my internship, and periodically when we pushed the button of the elevator, it would trigger a grand mal seizure. I'd seen it happen before, and it was the only explanation I could come up with for the fact that the elevator was not moving.

I glanced down at the bloody stretcher, and immediately ran up the stairs from the basement to the tenth floor, never stopping once to catch my breath. What kept me moving was the searing image of the unconscious Brinks guard bleeding on a stretcher in the basement, surrounded by rats, waiting for me to return. Finally reaching the top, I wrenched open the elevator doors to find Vinnie in his seat, staring blankly, like a zombie. I commandeered the controls of the elevator and rode down to the basement with a postictal Vinnie. I opened the doors to find the stretcher thankfully still there and no rats in sight. I jammed my foot against the door to make sure Vinnie couldn't take off without us and wheeled the stretcher in. The ride back up with the bleeding Brinks guard seemed to take forever, much longer than my sprint up the stairs. The old elevator rattled as it ascended, and I was sure it would give out, leaving me trapped helplessly with one dying man and one recovering epileptic.

It didn't give out. It got us back up to the tenth floor. I got the Brinks guard out of the elevator and into the operating room, which was supervised that night by Bennie Bub, a neurosurgeon from South Africa. I assisted with an immediate craniotomy. But in spite of the fact that we transfused unit after unit of donor blood into the patient, Dr. Bub just couldn't get control of the carotid artery tear. It felt like we were bailing water from a boat as it sank faster and faster.

I remember looking across the operating table at Bennie. There must have been an unspoken question in my eyes, one he answered with a sharp command: "Suck on the patty." I grabbed a suction tube to place on the surgical sponge and clear his visual field of the bleeding carotid. I knew then that we probably weren't going to save the patient, but we kept going. I'm not sure for how long, but it felt like many hours. We kept going until the patient died.

Whenever I told this story in years to come, I felt myself transported back into that ancient elevator with that dying man and incapacitated elevator operator, wondering if I had only been able to make it up the stairs faster—or if the elevator had worked properly, or more suction had been available, or blood had been transfused more quickly—would the Brinks guard have lived?

That was the first of many internship experiences to shape my approach as a physician, learning to keep my wits about me especially in trauma and emergency situations, and would remain with me long after they were over, and help me as I would provide critical care across the world.

There were similarly formative experiences in pediatric surgery with Dr. McDermott. Many little wounded children came to us needing all kinds of repair. One, an infant, arrived after his mother had put him into a near-boiling bathtub. He was badly burned, needed skin grafts, and we had to change his wounds daily. He wrapped his tiny hand around my index finger, and squeezed tighter than I thought an infant could. This seemed to calm him slightly. That grip, that little vise-like grip, became a familiar sensation as I anesthetized children through my training period—and I would feel it again as a new father, twice, when my newborn children wrapped their hands around my fingers, looking for something to tether themselves to in their new existence.

Another case that stays in my memory was that of a child who plummeted four stories from the roof of an elevator into the shaft of a housing project in Boston. As we repaired his nasty leg and arm fractures, I was overcome with disbelief that any child would have been left so unattended as to have gotten himself into such an extremely dangerous predicament.

Finally, I was deeply affected by my gynecological work, specifically treating the women, mostly young teenagers, who came in suffering from the after-effects of home abortions gone wrong. When they entered the hospital, they were in hell: lost, alone, full of fear and terrible abdominal pain. Most needed hysterectomies. I do not remember ever seeing a young man show up with them. Sometimes a friend, a sister, occasionally a mother. But never the man or the boy involved. The women were mostly suffering with abdominal sepsis. To think of those kinds of things going on in the unsanitary back-alleys of Boston, with coat hangers and without pain medication, was more than I could take. But I did learn to take it, and I would go on to volunteer for this kind of work for years.

Most of the surgeons, doctors and nursing staff at BCH had a terrific bedside manner. For the women and girls who came in suffering the effects of illegal abortions, this compassion was a vital part of their recovery, every bit as important as whatever surgery and drugs might be administered. They were surely suffering emotional trauma, including, for many, the pain of having to let go of any future dreams of motherhood.

Part of our job on the ward was to comfort these women. Sepsis polluted their blood, and caused painful suffering. We could administer antibiotics to combat that suffering, but there wasn't much we could do to speak to their minds and hearts in a way that might soothe them. We didn't have much in the way of psychological training. Mostly, it was just a matter of spending time with them, not judging them. Listening to and talking with them. Reassuring them that tomorrow would come, that they would heal, that this moment would fade into the past. Today doctors and medical institutions are much better equipped to anticipate and deal with psychological issues and post-traumatic stress. In those days, we just had to rely on our shared sense of humanity, on our bedside manner. Luckily, the U.S. Supreme Court allowed legal, sterile abortion in 1973, with the Roe vs Wade decision. I often worried that a future court, misguided and politically blinkered, would erode that groundbreaking decision.

These were my baptisms by fire. All doctors experience them. How they experience them depends on their own character, but also the character of their mentors and superiors. Mine—McDermott, Folkman, Norman, and others—taught not only skill, but respect, care, and attention. As it turned out, my medical career would follow a winding path that would take me to the field of anesthesia and intensive care of critically ill patients. They were often intubated and ventilated, which obviated the need for a great bedside manner. This is not to say that I have less than enormous respect for my patients, or for the doctors who perfect these skills. In many areas of medicine, these may be the only skills that really matter.

CHAPTER 6

Concentration, Camp

* * *

IN A DOCTOR'S LIFE, MANY stories concern death. Mortality—the immovable fact of it, how it can be avoided in certain narrow circumstances with the right combination of science and sense—is a dominant theme in a medical life. But it may be that the people who enter that life were already thinking about such matters. For me, I know that life and death occupied my consciousness long before I lost that gunshot victim at BCH, even before I wondered about my cadaver in medical school, and before I woke up in the hospital in Rawalpindi. As a young boy, I distinctly remember the moment I became aware that I was growing up among people who had stared death in the face.

I was visiting my grandfather Harry's barber shop on Blake Avenue at lunch. My grandmother Gussie took me. To get there, we traveled the ten blocks from home to Blake Avenue: an epic adventure for a child. Street hawkers towered overhead, their chests puffed up and their chins jutting out as they called out in Yiddish to the buyers passing by. I helped my *bubbe* push the cart filled with Harry's cooked soup past the stores, wiggling and snaking our way through the welter of humanity, and stopping to select the perfect pickle for my grandpa and me. The smells from delicious buns in bakery shops made my mouth water. The stink of dead chickens was less appetizing. They were plucked on the spot as we watched. White, downy feathers floated through the air like permanent snowflakes.

When, finally, we arrived at my *zadie's* shop, he greeted me cheerfully, delighted to receive provisions—not only the soup and the sour pickle, but his schnapps in a special flask. Gussie went to shop for food and left me to stay with the guys. The leather-upholstered barber chairs were for full-grown men. There was a full-grown man in the shop, not yet in the barber chair, waiting for his shave and haircut, reading the daily *Forward* newspaper. I sat down on a stool next to my grandpa, who offered me a bite of pickle.

Then there was an unfamiliar ritual: a toast. My *zadie* poured me half a thimbleful of fruity schnapps, gave a few ounces to the waiting customer, and led the toast. I gagged and coughed, my throat on fire. It tasted awful. *Zadie* Harry laughed, unconcerned. "Good for digestion!"

The customer dropped his newspaper and slapped my back. "Oy! *Boychik!*" he exclaimed in a raspy voice. "Now you join the society of men!"

I managed a smile, looking at him in his dark eyes with bluish circles under them. The man was only a few years older than my father, but he seemed aged, sitting there in his undershirt. My eyes fixed on the forearm that had just slapped my back, where I saw a string of six numbers crudely written in blue ink. What were these numbers? Why were they so important to the man that he would write them on his arm? But did he write them? They were facing away from him, meant to be read from the other side.

I turned to my grandpa. "Zadie?" I said, pointing to the man's forearm.

My grandpa covered my pointing finger with his fist. His grip tightened, a signal for me to be quiet. I clammed up.

He cleaned up the food, put on his barber's apron, and seated the man in the barber chair. The man sat back and closed his eyes. "We are all men here now," he said in his thick accent. He opened his eyes and pointed to the tattoo. "You want to know what is this?"

I didn't answer. My grandfather stayed silent.

"From Auschwitz," the man said.

I had heard that word before. It was usually said in hushed tones among adults. But I didn't really know what it meant.

The man flexed his arm muscle. "This number became my name. My work number. Everyone called me this." He pinched his tattoo, and the badly formed numbers bulged on his skin. I imagined a prison, or worse. He looked at me, a sad, empty look in his eyes. I looked at the floor. We all fell silent. I could hear my grandfather clipping hair. I saw clumps fall to the floor. I felt the lingering schnapps burn in my throat. I didn't know what to say. There was nothing to say.

✳ ✳ ✳

My maternal grandfather, Harry, was a barber who lived to 101.

Harry had arrived in New York long before the horrors described by the man in the barber chair. In 1903, he took a ship from Germany to Boston, didn't like the looks of it, and headed down to New York instead. My grandma Gussie had arrived in the city a few years before him. She came from Galicia at fifteen with her mother and got work at the Triangle Shirtwaist Factory. Crowded with other Jews and Italian women, she labored on the floor of the factory, locked inside during working hours seven days a week, sewing women's shirts in infamously awful conditions. But then she met Harry, a blonde-haired, blue-eyed buck. He promised to marry and take care of her. He was a barber with a good business. So she quit her job. Not long after, the Triangle Shirtwaist Factory caught fire. Gussie's workmates were trapped inside the locked doors, and 146 of her co-workers died or leapt to their death. She terrified me with this story and concluded with a warning: "They will work you to death if they can. Don't let them lock you in."

Later on, my grandmother almost died from a miscarriage. My mother remembered it clearly. The family was living on the Lower East Side. Her mother was bleeding and bleeding during her pregnancy, and instead of going to a midwife, she went to what is now New York University's Langone Hospital. There in 1910, she received one of the first human blood transfusions.

My father had also narrowly escaped death. He was in Germany at the rise of the Third Reich, although the path that took him there was tangled and strange, almost winding the wrong way through history. Ben grew up a poor immigrant from Russia. He moved to the U.S. in 1922, at the age of 12, but he was a smart guy, and he managed to graduate from an American high school and then from college at Long Island University. It may not have been a prestigious school, but it was a good local school, convenient and economical. He got a good basic education, and dreamed of becoming a doctor, a surgeon. Within a span of a few years, a series of tragedies befell him. His mother died of bone sarcoma, and his father passed away the following year. His brother Harry shot and killed himself because he had tuberculosis, and my father discovered the body. Plagued by tragedy, my father fell into a depression.

To escape the city and the sadness, Ben got a job working as a waiter at a summer resort in the Catskills, where the Jews in New York, New Jersey, and Philadelphia went to get out of the city heat. My mother was there, playing piano in a girls' music band. I've seen a picture of her band of four—fashionable and leggy ladies, like the *Marvelous Mrs. Maisel* of an earlier generation. He was lovestruck, and Florence was charmed. They fell in love and married in 1931. He still dreamed of medical school, but he couldn't afford it, even when he went from relative to relative begging for funds. So, in 1932, my father went to Germany. Yes, that's right: A young, newly married American man of Russian Jewish stock voluntarily left New York and relocated to Germany. He had gotten it into his head to enroll in the excellent, affordable medical school at the University of Bonn in Germany, where he figured he could pay tuition and maybe even pick up a little extra money by teaching English to Germans. He started his schooling there and studied under the famous Bonn anatomist Johannes Sobotta—the author of the anatomy textbook I would use in medical school. And he didn't just study under him; Sobotta also gave him a job as an assistant. This was during the time of the Weimar Republic, and Germany was still regarded as a sophisticated country, culturally progressive, medically advanced. My father was an avid German opera fan, spending hours in reverie over Wagner's *Der Ring des Nibelungen* (the Ring cycle).

My father was midway through his studies in Bonn when the Nazis transformed Weimar Germany into the Third Reich. The Nazi takeover in 1933 shocked the country and the world. The Nazis claimed they won a democratic election, but as we now know, that's simply not true. Conservatives in positions of power chose them over the communists as the "lesser of two evils."

Ben rarely spoke of that time. I think it baffled him. He didn't see it coming at all. Everyone he knew at the University felt the same as he did about the Nazis—that they were a bunch of awful hooligans. Still, they rose to power, and the University of Bonn was forced to expel Jewish students and teachers. In 1933, Hitler passed a law that foreigners like my father could not work in Germany. Sobotta's wife gave him dinner rolls to stuff his pockets every time she saw him, he said. Not enough to live by. My father's most cherished dream was to become a surgeon, and the fascists destroyed it. It was with a heavy heart that he had to say goodbye to his fellow students, many of whom would be drafted and then radicalized under the Nazi regime. I wonder: Did they ever think of Ben Zapol, their American Jewish friend of Russian extraction? Or were they so brainwashed by the Nazi Reich that such thoughts couldn't seep through? One thing was for sure, though, my father wasn't a bitter man. Years later he took me to the Metropolitan Opera to hear Wagner's Ring cycle. I could hear my father humming along with the orchestra. Still, the fact he had missed what he felt was his first and truest calling forever remained a source of sadness for him, and certainly influenced my parents' goals for me.

My father was one of those people who thrived in the face of adversity. He returned to America in February 1933, and did everything he could to stay alive during the Depression. He bought gold teeth from people who were selling their teeth, and for the gold, he could get a little money. He went into dress manufacturing with no knowledge and did okay, faking his way through. He then joined a questionable business started by his relatives illegally importing Swiss watch movements. When the corrupt nature of this business made Florence uneasy, he established a business selling Captain Marvel and Mary Marvel watches. It was this, I think, that made him lots of money. By 1950 this man who had been run out of Nazi Germany was a millionaire who drove a sleek Cadillac.

That's when he created Camp Zapol, a haven for middle-class Jews. I first started going to the area around Bloomingburg, New York, when I was an infant and World War II was in full swing. My parents had acquired property there during the Depression, advised to invest in land by Ben's mother. If there was something they didn't trust more than the banks, it was the stock market, especially then. Property, on the other hand, was a safe bet.

Their first investment was in a small house on the Roosa Gap Road, and that was the start of what eventually became Zapol's Cottages. Bloomingburg is in upstate New York, in the Shawangunk Mountains, part of the Catskills—where my parents first fell in love. The Catskills were a popular getaway for New Yorkers in the summertime, they are again now. I liked it there at any time of the year. We went in the dead of winter sometimes, when the snow would form in big drifts, silencing the entire landscape.

Still, this was the era before air-conditioning, so the area really came into its own in summer. If you had to stay in the city on hot summer days, it meant you had to live with the sweltering weather. Those that could get out of town did. They were well-to-do middle-class folk from New York. Jews weren't welcome at many of the summer colonies in the area. But they certainly were at Zapol's.

Zapol's Cottages turned into quite an investment, and my father added more and more cottages and attractions to bring people there. Small cottages sprouted on the land—some were shacks moved from elsewhere, and some he built new. By the late 1950s, Zapol's Cottages consisted of some 52 cottages on 250 acres of land. Eventually, in the 1950s, Zapol's cottages became Camp Zapol. We kids spent a lot of time outside, largely unsupervised. One thing we did, as much as possible, was watch movies. The best movies were made in that era. My father received reels on 16 millimeter film every week in the mail, and everybody would gather to watch them. Ida Rich made popcorn and her son, my best friend Marty, gave it out. *African Queen* with Humphrey Bogart has been running in my head ever since.

The beginning of each summer was like the same wonderful movie running over and over again. Kids would arrive and pour out of the cars, like so many fish suddenly let loose from a net. It was a terrific sight, like an animal migration. I'd be sitting on this one swing in the orchard waiting for my best friends and their families. The place was arranged on the basis of self-catering units, *kochaleyns*, meaning "cook alones" in Yiddish. Each family took care of its own breakfast and dinner, which let them get together in little groups and *schmooze*. You could also find them gathered around square tables playing Mahjong and Canasta.

It was up at the cottages that I had my first real exposure to the arts. My mother brought theater and music to Camp Zapol. She was constantly playing her grand piano. She played with great skill and loved to sing. She would take on students over the summer, drilling them on their scales like she did with me and my violin, and strongly encourage their participation in musical productions. Some that she directed are still classics, like Rodgers and Hammerstein's *Oklahoma* and *South Pacific*. Gilbert and Sullivan's *The Mikado* and *HMS Pinafore* to this day trigger my memory's reservoir of their unmatched clever lyrics and rapid tunes:

I am the very model of the modern major general
I've information animal, vegetable and mineral
I know the kings of England and all the fights historical
From Marathon to Waterloo in order categorical!

Aside from my one time playing the General, my role was playing the violin with my mother on the piano, or as an actor with a bit-part.

During the day, Zapol's Day Camp kept the kids entertained. About a half-mile from the *kochaleyns* was a swimming pool. Of course, there was a baseball field, too, but as a not-so-skilled athlete, my highlight on hot summer days without doubt was the swimming pool. I loved to swim. I started as a camper, and by the time I was fifteen years old I was a junior camp counselor, which meant I could teach other kids to swim. It was a role that suited me; I enjoyed helping the other kids, and it gave me a lot of confidence in myself.

Later in my life, while a student at MIT, I worked as a camp counselor again, which gave me the chance to make use of the experience I had earned at Camp Zapol, this time at Surprise Lake near Poughkeepsie. It was there that I met Joe Silk, a Cambridge University undergraduate from England—he was a co-counselor along with me, a serendipitous connection. At Surprise Lake, there were campers barely younger than me who had far more street sense, and were, shall we say, a bit rough around the edges. I remember watching them splay their hands out on the wood floor and throw knives between their fingers. I couldn't stop them. They were carrying weapons! But I'm glad we didn't have any unnecessary amputations and we all came out alive. Well, almost no complications. I remember one hike with campers when a sweet young boy stumbled, screamed out: "Snake!" and then screamed again. I rushed to his side and saw, so clearly, two deep punctures where the snake had planted its fangs in the kid's leg. Poison. I felt woozy, but I didn't faint. Instead, I lifted the child and carried him back down the hike, and out to a nearby farm, yelling for help. A bearded man in a truck with an Asian woman in the passenger seat pulled up. The child was wailing. I squinted at the man, who looked familiar.

"A hospital for a snake bite antidote, please!"

"Get in the back!"

He helped me into the back of the truck.

"Are you…?"

"Pete Seeger, yes, and this is my wife Toshi. We better hurry."

We made it to the hospital just in time. One of the greatest folksingers in history saved the boy's life.

When you peel back the layers, my counselor experiences partially explain the trajectory of my later life, my ongoing interest in the possibility of leading young people. When I strip everything back, that's what stands out and matters most; working in teams with people—of any age—and inspiring them.

Back to my father, and his financial prospects at Camp Zapol. While he was focused intensely on succeeding in his business ventures, my mother worried about the bottom line. Did we have enough money to keep the whole enterprise afloat? Would it last? Someone reading this might laugh at this. He drove a Cadillac, after all. And it's true that my parents had done well. But they hadn't come from much, and as Depression-era kids, constantly worrying about money was a matter of instinct. My father did his part at being frugal. He loved doing things himself. He took every task very seriously, whether he was hiring counselors or working with plumbers. However, because he (like me) didn't like to sit behind a desk, he often did the woodwork and plumbing himself. And you can bet, with 52 families occupying the cottages, the services of a plumber are often necessary.

Zapol's Cottages is one of the deepest anchors I have to my youth and childhood, not only for the experiences I had there but for the ways in which those experiences connected me to other kids. I had a biological brother, but I didn't even know he existed until I was thirty-six years old. My closest childhood friends, apart from my neighborhood pals and a few cousins, were my camp buddies Marty Rich and Stan Green, and his sister Linda (the camp sweetheart). Stan's father was an early convert to documenting family adventures with home movies, and his jumpy footage makes life at Zapol's Cottages come back alive. The memories of those summers anchor my childhood, and Sam Green's movies of those happy summer days are part of the sea floor.

The camp that I had heard about in Harry's barbershop and the camp that I had grown up in joyfully in the Catskills were worlds apart. One was history, the other family. But after the Second World War, the two converged in strange ways. In 1946, after the war was over, my father went back to Europe to look for missing relatives in displaced persons camps. He didn't find any, as far as I know, but that whole period was thick with a sense of what we had collectively lost. My aunts and uncles would whisper about these missing and dead at family gatherings, each year with increasing worry, even as the war receded into the past. Imagine my surprise when, at one such holiday dinner at my Aunt Sonya's house, a family who were truly long-lost relatives showed up. I was nine years old, and overjoyed to meet my college-aged cousin Eve and her parents, refugees from Europe who had survived the war in hiding and had resettled in Canada. Paula, Eve's mom, was an immensely gifted storyteller, a polylinguist. Her dad Elliot, was a chemist, a man of few words but great wisdom. They were incredible people who lit up my imagination.

Paula served my need for narrative. I peppered her with questions. "What was life back in Europe, in the old country?" She was pleased to hear my questions. Apparently other assimilated family members in America were unconcerned with the past. Paula told me about the Pale, where she grew up: an area that was sometimes Russia, sometimes Poland, confined to Jews. But as an "orphan" herself, whose mother died when she was four, she would wander across the tracks to where Jews were not welcome. She learned Polish and was able to pass. "If the world doesn't care about me, I'll make my own way," she said. It was a philosophy that served her well during the war, when she and her family evaded capture.

Elliot spoke to the other side of my mind. I remember him giving me some matches and then opening his hand to reveal a sugar cube. "Light the sugar cube," he said.

It seemed easy. Light match, touch sugar, right? I tried, but the cube wouldn't burn. Embarrassed, I tried again, with another match. Try as I might, it wouldn't burn.

"I'll do it," Elliot said, and he just lit the sugar cube. I smelled caramelized sugar, and the cube melted to a puddle.

Astonished, I stared at him. "How'd you do it?"

"Watch," he said. He took a fresh sugar cube and rubbed it in an ashtray. The metals in the ash catalyzed the burning of sugar, so he could light the cube on fire. Magic. Science. To me, they were the same.

In 1948, when I was six, I remember the celebrations in my house: the foundation of Israel. Instantly, Harry Truman was a God: he was the first government leader to recognize this new country. There were Israeli flags everywhere in my neighborhood. Israel, we were told, was a place the Jews could call home, a safe haven away from the bloodshed and terror that had swept through Europe. My parents dreamed of visiting. But they were proud to be Americans, comfortable in our Jewish home in Brooklyn.

That Jewish home included a fair dose of religion, but I did not enjoy my Judaic studies. To prepare for my Bar Mitzvah—a coming-of-age ceremony that occurs when a Jewish boy turns thirteen and is declared a man—I was forced to go to Talmud Torah four days a week at a Hebrew school down the street from me. It was located in the basement of a synagogue. A bunch of rabbis led me through memorization and prayer. I had no idea what I was saying, or why I was saying it.

I was so bored with it all, in fact, that I set up my first mouse experiment. I'd eat poppy seed bagels and put the seeds in the holes in the floor. The next day I'd check to see if the seeds were there, and if they were gone, I'd put more in there, for the mice. Each day, in the middle of prayers, I'd peer in the holes and replenish the seed, until one day, WHACK! I felt a sharp slap on the back of my head. I turned and saw Rabbi Pernicoff. Mean, mean man. That was the end of my first animal experiment—he made sure of it—and I was seated in the front of the classroom from then on.

The day of my Bar Mitzvah arrived. I remember it well, though not through my own recall—there is an old black-and-white movie that was taken of the festivities. When I look at it, I barely recognize myself; a rotund, pubescent boy, going through the motions required for the extravagant party. I wanted to make my parents proud, to earn a fountain pen and other gifts. But I am not sure that I had any sense of what it meant to become a man in the Jewish faith. My deeper connection to Judaism came later, when my wife and I helped establish a new synagogue in our town of Concord, Massachusetts, and when our two children and granddaughter had their Bar and Bat Mitzvahs. In that black-and-white footage, I can sense that the roots were being put down, but nothing had grown yet.

I sometimes wonder about the way in which it did grow. Both of my children married people who were not Jewish. When they each started dating their eventual partners, I was upset and fearful. What would happen to our traditions? Why were they rejecting our culture, assimilating to the point of erasure? Deep down, I feared that I hadn't raised them to respect their Judaism, or that they were doing this all to spite me.

Nikki asked me why I was behaving in such a tribal manner. Couldn't I see that our children had chosen wonderfully? I could, I told her, but I didn't want to see our traditions disappear. I reflected back to a time in my early twenties when my mother found out that I was dating a gentile. To say that she was upset is an understatement. She would look at me with icy stares. Our phone conversations changed from easy discussion to vast and uncomfortable silences. The relationship with that girlfriend fell apart naturally, but afterwards I put in a little extra effort to find a nice Jewish girl I could bring home.

Ironically, Nikki Kaplan—the nice Jewish girl I did eventually bring home—wasn't an ideal candidate in that regard. Her parents weren't observant Jews, she hadn't had a Bat Mitzvah and didn't know much at all about Judaism or Jewish traditions, and Nikki certainly wasn't a Jewish name. But after my parents met her, and her parents, she passed their test anyway. I suppose they decided she was sufficiently Jewish, and they liked her! (What was not to like?) Once we had children, I felt quite strongly that they should have some Jewish education to learn about their heritage. Mercifully, their Jewish education was better than mine—Rabbi Pernicoff was long gone. That was part of why I was so confused when they both found partners who didn't share our faith.

Over time, these worries faded. Love came in instead. I love Diana, my daughter-in-law, and DT, my son-in-law. And I also love seeing how my own children have kept connected to their cultural and religious identities. David studied Yiddish, and he spoke so movingly and vividly about the great Yiddish literature that it inspired me to take up Yiddish learning too. Liza has done several oral history projects about Jewish New Yorkers, from Holocaust survivors to multigenerational matzo factory owners. And all of my three grandchildren, Ruthie, Elliot, and Juno, are learning about Judaism, celebrate Jewish holidays, and most importantly are healthy, funny, smart, loved children. I couldn't be prouder of them all. As I write this, I think about my granddaughter Ruthie's Bat Mitzvah in San Francisco. The beautiful ceremony, in English and Hebrew, connected her present to my past, carrying us both into the future.

With Nikki and our grandchildren in Miami Beach in 2018.
From left, in the mango tree: Ruthie, Juno and Elliot.

CHAPTER 7

Lightning in a Bottle

∗ ∗ ∗

WHEN I WAS A KID at Camp Zapol, my radio would break down and I would have to open up the cover and stick my fingers into my old Heathkit rig to change a tube. That would invariably lead to big arcs of electricity lighting up the night. I grew to enjoy the smell of ozone from sparks in the air, albeit sometimes mixed with the smell of burning hair. It came to signify progress in engineering for me. When it would get late, I'd walk out into the humid midnight air in Bloomingburg and watch the fireflies pulsing in the night. Later in life, I'd run around with my kids and a butterfly net in Concord, Massachusetts, trying to capture one of those magical glowing summer creatures. My fascination with electricity and light continued for my entire life.

I'm a scientist first, but I'm also an inventor. Sometimes when we invent, the invention is like a firefly. It burns bright for a time, and then perhaps it dims. It's not always clear at first what we have. As my old teacher Isaac Asimov once said, "It is only afterward that a new idea seems reasonable. To begin with, it usually seems unreasonable." What stops that dimming, that fading away? It can come back to light if we revisit the idea, or someone else takes a hold of it. Over time, if we're lucky, that pulsing light grows and multiplies, and then there are many glowing lights, which together form a constant source of light. Inventions can illuminate and make the world better. That's why I invent. I am a doctor, after all. I have sworn the Hippocratic oath to do no harm. Sometimes the only way to do no harm is to find a new way to do good.

In the 1990s, as we continued our work showing that inhalation of NO could save babies in the clinic, I learned that the U.S. Army had flown fighter planes through lightning storms and had measured gasses. I think they were interested in ozone creation because we were newly aware at that time of the ozone hole over Antarctica. At McMurdo Sound in 1992, I had heard a talk by the charming Italian scientists who were sending weather balloons and shining lasers up into the sky to make those measurements. I suppose the Army wanted to get in on the measurement game as well, so they skipped the balloon, sent some daredevil pilot right into a lightning storm, and the instruments they carried measured high levels of NO after the lightning struck. The lightbulb went on for me when I read this. "Aha," I said. "Lightning makes nitric oxide…. I know how to make lightning."

The smell of ozone returned to me then, memories of Bloomingburg. Had I been inhaling nitric oxide as a kid? We designed an experiment to find out. We started with simple devices: clickers that you light a gas barbeque with, little piezo-electric devices where you squeeze a crystal and it sends out a pulse of electricity. We had very expensive measuring devices, and we could see that it would take a lot of clicking to make a significant amount of NO, so we dug in at the laboratory and constructed a Frankenstein-like machine to generate big pulses of electricity. Kevin Stanek, a friendly, round-faced, quiet engineer, gave up half of his office to one of our creations. It had dials and switches and meters and looked like one of my old ham radios. It could produce a big bang and send out NO and ozone. I remember the radio playing, the Red Hot Chili Peppers or some other group, and then there was the loud bang and whiffs of ozone. We patented our contraption, figuring that with some refinement, this could be a way to get NO out into the world.

It was, in that sense, a medicinal delivery system. We knew by then NO was a medicine. We knew its benefits for humanity in general, for babies in particular. But how could we deliver that medicine to the many, possibly millions, of people whose life it could save? Those first stunningly successful tests performed on babies who turned from blue to pink were done with a cumbersome, complicated combination of heavy cylinders of NO, a common industrial gas, and devices we had built to control the flow of the gas. To make it look benign, we topped it all off with a coffee box with a face drawn on it. This was clearly not going to be everyone's idea of a practical medicinal therapy, so in my mind it was time to go get this into the hands of people who could do more to develop our idea and pair it with the right technology. I knew that our team and our NIH grants would only take us so far, and I would have to find deep pockets to make this life-saving gas available worldwide. Electrically generated NO will be the future, I have no doubt, but our search for those deep pockets led us in other directions first.

I had some history when it came to finding and emptying pockets. My dad, remember, was a salesman. He could talk to anyone and make them smile, and quite often they'd buy into his scheme. In his days selling watches, he had even had the chutzpah to file a patent of his own (U.S. patent USD163729S: A watch commemorating the Pope's 1950 visit to the United States.) It was not one of his greatest ventures, but he was never deterred, and at some level I must have absorbed some of that ability to sell, and not to be too affected by failure. This is an essential characteristic of a research scientist. All the good ones I've ever known are relentless, and the best are stubborn as goats.

I knew I had to be like my dad and go out and pound the pavement to get companies interested. I had no heartfelt interest in the business side of this, but I was interested in doing clinical trials and treating patients, in making NO available around the world. I also had great pride in what we had done and wanted to see my fireflies shine.

Charlotte Harrison and Marvin Guthrie in the MGH Office of Technology Affairs backed me up with the appropriate legal support, and Claes Frostell and I made many pitches to many companies. We knew our opening line: "We can save babies with NO." The first wave of responses wasn't promising.

"We can save babies with NO."

"Oh really, that is a gas, right, not a pill. We don't do gasses, we only make pills."

"We can save babies with NO."

"Oh really, that is a drug, right? We only do gasses, we don't know about drugs."

"We can save babies with NO."

"Oh really, you can save babies? We don't develop drugs for babies. Too risky."

"We can save babies with NO."

"You've only treated a few dozen. We'll need more proof."

This feedback eventually made me realize we just might have a chance with companies that sold anesthetic vapors. They might understand the intersection of gas and medicine. I found a connection to Anaquest, Inc. through Keith Miller, another inventor in the MGH Department of Anesthesia.

Anaquest was the pharmaceutical division of the British Oxygen Company (BOC). In the past, BOC had produced, distributed and sold anesthetic vapors such as isoflurane, so they were familiar with the lengthy and rigorous process of taking a product through trials and getting approval from the FDA. This meant that they were also familiar with the rewards of doing so. BOC had a large gas business, their main products being bulk commercial gasses like oxygen, carbon dioxide and so on, so they had the necessary industrial muscle and know-how for the job. I went to Eve Jelstrom, a Certified Registered Nurse there, and started the conversation with her the same as I had with the others. "We can save babies with NO." I braced myself for the rejection that never came. "That is amazing!" Eve said. "Let me assess the market and I'll get back to you."

Eve was the first to see the glow of promise in our research, and—as promised—she set out to ask various experts in their fields to weigh-in on the results of our trials, including anesthesiologists, pediatricians and neonatologists. She was able to convince her company to make a grant to fund a small project with us. I can remember her telling me about describing our work to Dr. Mark Rogers, then Anesthetist-in-Chief at Johns Hopkins. "Well," Dr. Rogers said, "if a selective pulmonary vasodilator is found, that inventor should get a Nobel Prize." Mark saw not only the glow, but the big light that was making it. And as it turned out, he was not far off the mark in his Nobel prediction.

After a few months, Eve and her colleagues sent a sharp young Englishman named Ashleigh Palmer to our lab. Ashleigh was dressed well, with cufflinks and a colorful tie, and seemed quite comfortable in the lab despite being an MBA. I would later learn that he had studied biology as an undergraduate, and so he understood early-stage research and development. And in time he would confess to me that he wasn't sent there to observe our experiment. He was sent there to terminate our project with BOC. Before he could start in with his speech about the difficulties of continued funding, I posed a question to him: "Would you like to see a baby whose life we've just saved with nitric oxide?"

He was curious enough that I made a call to Jay Roberts, my friend the neonatologist. "Jay, would it be ok if we walked over to the NICU?"

"That would be fine, Warren," he said. "We're giving NO to a baby who was born yesterday, and she seems to be doing well."

When we arrived at the NICU, we put on surgical gowns and gloves and walked by the whirring ventilators and little incubators. The babies inside were unimaginably small babies. We came to the enormous cylinder of gas connected to a delivery contraption, our Frankenstein, towering over the tiny baby breathing NO in the incubator. Jay raised his gloved hand, waved a welcome to Ashleigh and explained what was happening: "She was born last night up in New Hampshire. She arrived by helicopter three hours ago with oxygen saturation in the 70s. We started NO right away. Within minutes her sats were in the 90s. We're going to keep her on this dose for a while until we're sure she is stable."

Something in Jay's description sparked something in Ashleigh. I could feel his excitement. Maybe it was his scientific background. Maybe it was another thing I didn't know at the time, which was that he and his wife were about to welcome their first child into the world. Maybe he could see his child in this child, or maybe he was thinking of all the children in the world. At any rate, something lit up in Ashleigh, and when we left the lab and rode back down to the lobby, he couldn't stop talking about the remarkable young girl who we'd just seen. There is nothing quite like seeing the miracle of a new medicine working on a patient, saving a baby, and believing that this is going to change the world. When we reached the front desk, he turned to me and gripped my hand hard. "This is truly amazing, Warren," he said. "I had no idea. I will see what I can do. I don't know if I will be successful, I'm in a big company with a mind of its own, but I'm going to give it everything I've got."

Everything he had turned out to be more than enough. With remarkable speed, Ashleigh convinced the once-reluctant company to pay for a nine-month option fee that would allow him to research and build a business plan, recruit a team and sketch out a clinical trial strategy. His argument was quite simple: that they could make a lot more money making a specialized patented medicine than they were making from similar gasses like oxygen, which were commodities in the hospital market. It was a straightforward economic argument, which would have never occurred to me. This strategy was successful, and BOC became our first industrial partner. As Asheigh liked to say, with his characteristic dry English humor, his failure to do his job and kill the project fortunately didn't get him fired. It was so much more than that: it led to our decades-long collaboration in which we worked together to bring NO to the world.

Meanwhile in Sweden, Claes had befriended Rolf Petersen at Aktiebolaget Gas Accumulator (AGA), a Swedish Gas company founded in 1904. They also became very interested in realizing the potential of NO. Working with AGA made sense. They could test NO in Europe, get regulatory approval and distribute it not only there but worldwide. If the FDA seemed daunting, I couldn't imagine having to deal with regulatory agencies in dozens or hundreds of different countries. Fortunately, Rolf understood the importance of our invention, and would be our champion in ways we could not yet see.

Ultimately Charlotte Harrison and Marvin Guthrie at MGH would negotiate licenses with both companies. Although it may seem unusual to license the same inventions to two large companies, in this case it made sense. Each had their strengths: BOC had the pharmaceutical skills and U.S. presence that were so valuable for collaborating with us on this project; and AGA had its distribution systems. The two got along well. It wasn't a complete success, in part because neither of them fully understood the power of electricity. As part of the licenses, the companies gained control of our patent on electrically generated NO. But as gas companies, they focused on making tank-based businesses. That kept the firefly—the promise of widely available electrically-generated NO—in the jar on the shelf. I would need to learn to push that part of the process forward myself. Decades later, I would urge the hospital to write a letter to the company that acquired the NO business, asking them to relinquish that patent.

Over the years, AGA and BOC would send me toy trucks in the mail. They were models of the kind of milk trucks or gas tankers you see on the highway, big long cylinders on long trailers, towed by a driver in a cab. These were fun for my grandkids to play with, and were a nice concrete illustration of what effort they took to ship gas around the world. They were lumbering gas companies with big factories that bottled gas and trucked it across the world. BOC ultimately constructed a $80 million chemical synthesis plant in Louisiana to produce NO gas stored in nitrogen for inhalation. It was one of the most gratifying days of my professional life when I visited that plant with the president of the business, Roger Stoll. It must have been in the mid-1990s when Roger proudly told me that more than 200 U.S. jobs were assured for NO production for many years if NO were approved by the FDA as a baby therapy. I was struck, not just by the Louisiana humidity and heat, but by how amazing it was that all this came from my MGH laboratory, years of hard team work and NIH grants. I was also struck by how arcane the gas business was even then, envisioning the day when we could get rid of that factory and make NO with electricity.

In order to ensure those jobs in Louisiana, and to treat babies in the United States, we had to gain approval from the FDA to sell the gas. Ashleigh hired an old, bushy bearded pediatrician colleague of mine, Dr. Richard Straube, to run the clinical trials, and together we went to the FDA to find out what they would want from us.

Our first meeting had all the charm of an icy plunge into Antarctic waters. We were the last interview of the day in the FDA Cardiovascular and Renal Drug Products Division. The staff looked exhausted and as we walked into the enormous room, a few of them shook my hand as they departed, clocking off early. A few dedicated staff remained. I shook the hand of the lead reviewer, the venerable Ray Lipicky, an FDA scientist-regulator with deep roots in cardiovascular pharmacology. He held an appointment at Woods Hole Oceanographic Institute for 30 summers experimenting on drug effects in squid axons. I found his intellect to be also somewhat tentacular—extending into all sorts of unexpected domains—and he was extraordinarily gentle and Socratic, even when his lessons were icy cold. He was known for his principled approach and high standards for approval. I wasn't worried. I thought we were up to the intellectual challenge.

My motto at the time was "show me the data," so we got right down to it and presented the data that we had collected in animals and in human babies who had received treatment with NO. The data were clear, and much of them had been published in highly respected peer-reviewed journals. I told Ray I hoped he would approve the use of the drug based on the work we had done.

NO didn't cause illness in healthy babies at the concentrations we were using, and we'd seen miraculous recoveries time and time again. Ray pulled out the 1986 edition of Gosselin, Hodge et al, *Clinical Toxicology of Commercial Products*, and read to us the toxicologist's bleak view of what happens after exposure to gaseous NO:

> *Increasingly rapid and shallow respiration, cyanosis, mild or violent coughing with frothy expectoration, and physical signs of pulmonary edema… Anxiety, mental confusion, lethargy, and finally loss of consciousness.*

I felt a bit like a little boy who had been working all day on something for his mother, only to be met with derision and dismissal. Ray was not in the least bit impressed with our data, and he was seemingly impervious to the stories and lives already saved by this novel treatment. His response was to disregard our experiments—not to mention our industrial partners— and return to the conventional wisdom of the textbooks.

I was stunned, almost too much so to defend the idea properly. But I caught myself and forged ahead. "We have evidence of selective pulmonary vasodilation in a dozen sick infants—with unchanged systemic hemodynamic measurements and greatly improved oxygenation. What more will we need to be able to make this therapy available?" I asked.

Carefully, Ray laid it out. "First of all," he said, "we will need well-controlled, well-powered toxicology studies in several species." This would mean millions of dollars of studies, which our company partners would have to shoulder. There was no way the NIH would give us a grant for those.

"And we will need two double-blind randomized clinical trials in the patient population." This would mean a sophisticated clinical trial organization would need to come up with plans to enroll hundreds of patients, not dozens. I knew this would take years to complete, and cost hundreds of millions of dollars.

"And finally," he added, "You can count toes up, toes down. It doesn't take much technology and it's very convincing."

This was the kicker. This meant that he wanted us to prove mortality benefit. He wanted us to be clear about the fact that it saved lives. While this was a noble goal, the number of patients required and the length of trial required to do this would prove insurmountable. Ultimately, he and I both knew that we'd have to find some middle ground that was feasible. Without some flexibility the FDA would only succeed in squashing innovation, and Americans would never see new medicines.

Though not that day, Rick Straube, along with an army of regulatory consultants and advisors, ultimately convinced Ray to agree that if we measured a combination of a decrease in mortality and a decrease in the use of extracorporeal membrane oxygenation (ECMO), we would be providing measurable benefit. This was just our luck because Ray, as well as most of our country's neonatologists, never liked ECMO. And for good reason. Though I helped to develop that technology with Ted Kolobow at NIH and in Vietnam too, ECMO is far from ideal. While sometimes being the only treatment option, it is very invasive, expensive, time-consuming and risky, with 15% of heparinized babies on ECMO hemorrhaging into their brains.

In 1991, we published our first studies in lambs, which caused a great stir. A year later we published our studies confirming the potential of NO to turn blue babies pink. That same year, Drs. Steve Abman and John Kinsella in Denver, Colorado published the results of their independent research, confirming our results. This would in turn reduce the necessity for ECMO and save patients' lives. Very soon, neonatologists were clamoring for access to NO to treat hypoxic newborns. This was problematic to say the least, since no American company was at that time selling it to hospitals for human use. Compassionate use protocols were quickly drawn up and approved in hundreds of newborn ICUs around the USA and abroad, but NO gas supply was still a problem. We had gone from being looked upon as lunatics to having the thing that everyone wanted and couldn't get their hands on. It was a remarkable reversal. Inspired by Ashleigh's strategic genius, his company, BOC, started giving the NO gas away free of charge across the USA.

It was the beginning of a long road. It was to take Ashleigh and Rick a further eight years of five large, grueling randomized studies paid for by

BOC, AGA and the NIH to satisfy the FDA requirements. These large trials are well chronicled in the literature. There were numerous meetings at NIH where clinicians would clamor for NO to be approved, only to be disappointed. Feeling despondent after one of them, I had a chat with Dr. Lipicky. He probably sensed my mood, and to be fair, his advice was remarkably insightful. "Warren," he said, "Be patient, if we approve NO now you will never know for certain if it works, as there will never be another term-baby study if we approve it." As we all know, the possible applications went beyond babies, but sorting them out was not easy. There was hope that we would gain approval for adults with ARDS, but those trials were not successful in part because in those days we would ventilate patients with much higher pressures of air than we do today. We would effectively blow up their lungs. So while we may have helped those adults with NO, we definitely were hurting them with our ventilators. I have made the case many times since that those studies should be repeated with modern ventilation technology.

✳ ✳ ✳

Nothing was easy, ever. In 1997, after about a hundred million dollars had been invested in gas manufacturing plants and clinical trials, Ashleigh called me up. "BOC wants me to kill the project again," he said.

"What? Why? Is there bad data coming from the clinical trial?"

"No, no." he said. "BOC is going to sell its Ohmeda division to Baxter International." They had decided that they didn't want the medical gasses in the company anymore, and Baxter was the winning suitor.

"Oh, that is great," I said. My tone wasn't sarcastic, exactly, but it betrayed my anxiety. I had tried to convince Baxter to take on the NO project before, but I was out of my league as an academic scientist. I knew Ashleigh was operating in a different world.

"The investment bankers say Baxter will close the deal in 30 days, but only if we kill the NO project. They don't want the risk of killing babies in a neonatal trial."

Now I didn't know what to say, so I didn't say anything.

"I just want you to know that I'm still working on this," he said. "We're only a year or two away from approval, and the data still looks good for babies."

Now I had to speak. "What if they stop supporting the trial?"

"I don't know. I talked them into giving me the 30 days to come up with an alternative, and I've got an idea. Do you have Rolf Petersen's number?"

Rolf was at his family's summer house on the Stockholm Archipelago. Ashleigh reached him there, and the two agreed that they had to keep the NO project alive. Miraculously, over the next 30 days, they raised the support and funding required. I honestly have no idea how they did it, other than the combined sheer force of personality and conviction. They formed a new company, INOTherapeutics, and BOC agreed, for a small token payment, to transfer the MGH license into the new company, along with the ownership of the $80M NO plant in Louisiana.

The trials went on until 1998. During that time, I was busy with administrative duties, running the Department of Anesthesia at MGH, so I didn't pay close attention to what was happening in the lab or the company. The department was demanding, and then some: there were budgets to make, staff to mentor and grow, hospital dynamics to navigate. I would call Rick every once in a while to ask how things were coming, and ask why we weren't on the market yet. As we got to the end of the trials, the Data Safety and Monitoring Board unmasked the data. It seemed promising. As we reviewed the data Rick pulled me aside. "We're going to need a new PI," he said.

"What?" I said. "We're almost done with the trials, we have plenty of principal investigators." I was certain that Jay and his colleagues had been good researchers and done the best anyone could hope for.

"No," Rick said. "Not that kind of PI." I must have looked confused. "We are going to need someone to track down all the babies to make sure they are still alive. Remember Ray's request that we measure "toes up, toes down"? Every baby who has left the hospital needs to be counted. If we don't know if they are alive, they will count against us in the final tally."

"You mean you want to hire a *private investigator* to find the babies outside of the hospital? Is that even legal?" I was incredulous—and impressed.

That was exactly what Rick wanted to do, and exactly what he and Ashleigh did. By the end of the year we knew which babies had survived the

NO trials. Ashleigh told colorful stories of the PI crossing into Mexico to find a baby who had gone missing from the hospital records. This was essential for success, and in the end they brought the total to only 44 deaths across placebo and treatment arms in the two pivotal trials, or about 10% of all subjects enrolled. It's one of my favorite examples of our team doing whatever it took to get a job done.

In December 1999, after reaching a positive outcome in two double-blind, randomized, placebo-controlled, multicenter trials of inhaled NO in a total of 421 infants, the FDA finally approved inhaled NO as a drug for term and near-term infants with hypoxic respiratory failure. This great success, this crowning achievement was the culmination of the leaps of faith taken by hundreds of courageous parents and thousands of other people over many, many years. Finally, NO would see the light of day and its potential to save lives would be unleashed. Our supporters were ecstatic. After investing hundreds of millions of dollars developing and testing this novel therapy, AGA and INOTherapeutics were eager to begin sales.

The 1999 FDA approval remains one of the most defining and proudest moments of my career. It marked the end of a long journey that started with an off-the-cuff hypothesis about NO when I was visiting Lou Ignarro in California many years before. It passed through every phase imaginable: from a two-person team using tanks of NO with lambs in our modest MGH lab, to witnessing a human infant breathe NO and turn from blue to pink, to drumming up support from international industrial partners and finally—finally!—navigating the perilous process of FDA approval. Like so many journeys, it was one that could have been derailed at so many places along the way—if I had had second thoughts, or if key people like Ashleigh or Rolf hadn't faithfully stuck with it through thick and thin.

Now, however, came a really tough question: What would the price be? Ashleigh reasoned that since NO was effectively replacing ECMO as a therapy, the cost of ECMO treatment should set the price of NO. Ashleigh asked the University of Pittsburgh medical economics consulting group to help him with these calculations. He priced ECMO therapy at $50,000 per treatment, allowing INOTherapeutics to price gas therapy at $3,000 per day with a cap of $12,000. This would provide inhaled NO to an infant for up to a month.

The company delivered the whole therapeutic package: the delivery device, calibrated gas monitors, and pharmaceutical grade tanks of NO.

The community of neonatologists was stunned. A big part of it was sticker shock. They had little understanding or appreciation for the years of industry money that had been poured into the development and approval of NO as a drug. They had received free tanks of NO for the years during which it was experimental, to the benefit of their patients and reduction in the costs that would otherwise have been spent on ECMO. As hospital managers always have a hundred leaks to plug, they had allocated these cost savings elsewhere and had no budget capacity for NO therapy when it was later approved. And NO use was quickly exceeding that of ECMO, meaning that the overall spend on respiratory therapy was going up. So NO was not seen as a cost savings, but a price gouging. The U.S. market for NO rapidly topped $100M per year and went on to reach $300M per year and more. INOTherapeutics was bought and sold as a valuable asset, and competition jumped in when the patents expired. Even in 2020, the price remained so high that it was a serious impediment to using NO, despite reimbursement from insurance companies for the neonatal uses. From my clinical and academic perspective, it was a real learning curve, to put it mildly, to observe the contentious interface between research, industry, insurance and clinical practice where pricing is concerned.

Even after all that, my hope for making cheap effective NO available worldwide was still glowing in a jar on the shelf. I never lost sight of it. The time had come to make it a reality. My son David had been working with Pfizer on an antimalarial program in Kenya (he went on to help build the largest pharmacy chain in East Africa—Goodlife). He got me fired up again about using electric NO to decrease the cost, eliminate the supply chain, and build the availability of NO worldwide. That became my new goal when I was working in Mbarara, Uganda with Médecins Sans Frontières on a clinical trial of NO for severe cerebral malaria. My daughter Liza, an oral historian, also participated, masterminding a beautiful video that I commissioned to document those trials.

She worked with Jonathan Lowenstein, my dear friend Ed Lowenstein's son and a professional photographer.

Now that billions had poured into the coffers of large companies, it was time to address the serious access barriers. MGH had successfully recovered the patents, and in 2014 we finally launched a company, Third Pole, to make NO, *generated by electricity*, widely available and affordable.

David and I hit the pavement again. We were channeling the Zapol sales line again, reaching back into my father's career. We talked to investors, grantors, and big companies. We visited Linde, the company that had bought BOC, and went back to their board room, where we reminded them of the big tanker toys that they had sent me over the years and gave them a toy electric car. That was our way of illustrating that we meant to disrupt the industry.

The conversations we had about electrically generated NO sounded a lot like some of the conversations we'd had twenty years earlier, with a twist:

"We can save more babies with electric NO."

"Oh really, that is a device, right? We only do gas, we don't know about devices."

"We can save more babies with electric NO."

"Oh really, that is a drug, right? We only do devices, we don't know about drugs."

"We can save more babies with electric NO."

"Oh really, that is a gas, right? We only do drugs, we don't know about gasses."

You get the picture.

This time around, our first support came from NIH via the Boston Biomedical Innovation Center (B-BIC). We were lucky to recruit to our board both Ashleigh Palmer, who by then was in the process of building another billion dollar corporation from scratch, and Jeff McCormick, a Boston financier who had worked with David in Africa. Wolfgang Scholz, a veteran Siemens product developer who I had worked with to design the first ICU patient monitors, joined to lead product marketing. Our team won the support of Johnson and Johnson through a competition. The FDA even gave us a grant to help us get a new therapy on the market for kids.

There were cultural differences and expectations to navigate, as always.

When one Chinese company came to meet us in Boston, I had a special greeting prepared. "Thank you for coming to visit us," I said. "First, I want to be clear—David is my son. I hope this relationship does not concern you with regard to the future of our company or a potential partnership."

"Oh, no, not at all Dr. Zapol," the head of the delegation replied. "Have you ever heard of Johnson and Johnson?"

Over time we gained the support of world-class investors and hired a fantastic team and board who have navigated the ups and downs of our company since its founding. We took the jar down from the shelf, and like the fireflies in Bloomingburg I believe that electric NO now has a life of its own.

It remains my most fervent hope that in the future there will be more conditions for which inhaled NO will prove to be a safe and beneficial therapy. I believe it should be more widely adopted. More patients will be saved from disease if government, academia, and industry cooperate in a climate of mutual trust and close communication; if they fund and pursue the necessary science; and if champions in industry navigate the hurdles that are associated with bringing new medical treatments to market. Despite those initial misgivings about pricing, NO is cheaper, more effective and much, much less traumatic to the newborn than ECMO. Now that the MGH patent is long expired, other manufacturers like Praxair and Airgas are distributing FDA-approved NO, and reducing the cost. Third Pole is working hard, having raised over $100M to bring electric NO to market worldwide and radically transform the landscape.

That entire process remains central to my life—to understanding that inventions never exist in a vacuum, but that they must interact with real-world forces. They must navigate money, corporate cultures, and government procedures. I have taught entrepreneurship to Harvard Medical School/MIT Health Sciences and Technology students over the years, and even more than winning the Inventor of the Year prize for NO, I am proud of the teaching prize I won for that course. I often teach these stories from the history of the commercialization of inhaled NO in hopes that they will help the next generation of entrepreneurs in health sciences to keep the light alive. Being a stubborn goat is half the battle. The other half is keeping your footing as you go up the mountain.

CHAPTER 8
Tinkering

✱ ✱ ✱

I HAVE MET ONLY A FEW mad scientists like me—people who love to create contraptions, people who think outside the box, people who create new boxes. I've learned so much from these people, and I've noticed that, like me, each one continues to tinker in their area of interest over time. What you make can always be better, more efficient, less expensive, and far more accessible. The world eventually innovates and catches up to your ideas. Even so, I've had many ideas too early, and have had to wait for them to have currency. I don't like that, in that I don't like to wait at all.

In my transformation into a restless tinkerer, I learned the most from my experiences in the late 1960s and early 1970s. First, I was transformed by the magnificent scientist inventors I met at the National Institutes of Health (NIH). Scientist clinicians at NIH in the Public Health Service were called Yellow Berets, and in the summer of 1967 I interviewed to become one, specifically to become Senior Assistant Surgeon at the National Heart Institute at the NIH in Bethesda, Maryland. National competition for these jobs was high and attracted the best American academic medical graduates. A few went on to be awarded Nobel Prizes.

I felt my hands sweat and my heart race when I went for my interview. For men my age, if you didn't get a job, the most common alternative was serving in the military in Vietnam. The stakes were high. Greeting me for my interview was a thin, eggshell-bald man, who introduced himself as scientist-inventor-physician Robert L. Bowman. I was starstruck. Bowman was the Chief of the Laboratory for Technical Development, had been present at the founding of NIH, and was himself a star inventor. Bob had invented the Aminco-Bowman spectrophotometer, an optical chemistry instrument with two prisms for examining light in and out. The instrument was in use around the world to identify chemicals, and had gained Bob fame in a variety of scientific fields. He was also an editor of the preeminent journal *Science*.

I listened to Bob as he told me a bit about himself. Like me, he was from New York, but unlike me, he had no trace of an accent. He had served as a psychiatrist in the Pacific during World War II. (Imagine the stories he heard!) When he went to NYU medical school, he was trained in anesthesia by Emery A. Rovenstine, who established the first academic department of anesthesia at Bellevue Hospital. Then came the interview. It was long and stressful. At the conclusion, Bob looked up from his notes. He had a look on his face as if he was about to ask the most important question of all. "Where did you go to high school?"

I looked at him quizzically. "Stuyvesant."

He immediately broke into a broad smile. Something in the air palpably shifted. He shook my hand. "Me too. You're hired."

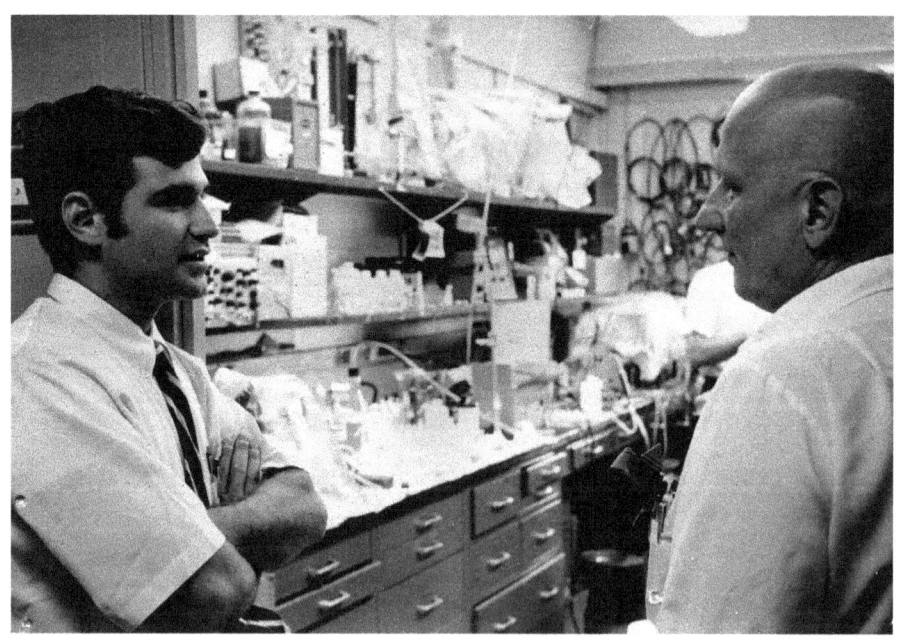

With Robert L. Bowman in his NIH laboratory, 1967.

I could scarcely believe it. But I had a feeling I was going to work for a man I could count on.

I soon met Ted Kolobow, a member of Bowman's laboratory. Ted was thin, with bright blue eyes that expressed much more than his lips. When I first met him, he was serious, rarely smiled, and spoke very little, even at important meetings. He advised me to also speak little, but to publish well and often—a lesson I may not always heed but have never forgotten! Ted was a medical genius who had invented a disposable spiral coil silicone membrane oxygenator while he was still a medical student at Western Reserve. As the first disposable oxygenator, it had also gained Ted renown in the medical world.

Though laconic in public settings, he loved to talk when we were alone. He first spilled his story at a restaurant after work, when I offered to split a beer. "No, no. I don't drink. I can't drink," he explained. He continued: He had lived through World War II, spending the latter part in a German displaced persons camp. Ted was a refugee from Estonia—one of the many countries occupied initially by the Soviets, only to be taken over by the Nazis, and he hated both. He escaped from Soviet Estonia into Nazi Germany, traveling over the border hidden in a horse cart filled with vodka. Ted could only hydrate with alcohol, which not only made him sick and drunk for the journey but put him off booze for the rest of his life. No beer at our lunches. He also had a habit of skipping main courses in favor of a double helping of dessert. He had the sweetest tooth I ever saw, but I never asked him about it. I knew from my upbringing that survivors of Nazi persecution were often troubled by dietary issues. It never seemed to impact his brilliant mind.

Over one of these lunches Ted, his bright eyes wide and vividly blue, told me an incredible story of how an eminent Professor at Western Reserve Medical School had attempted to claim inventorship of Ted's invention while Ted was away on summer vacation. My jaw dropped listening to this. The gall! But Ted told it like it was no big thing. That's how he told all his incredible stories, cheerfully putting sweets into his mouth and smiling as though this was just another twist of fate that he survived.

I went to work with Ted to help test his disposable spiral coil oxygenator in animals, a critical decision that determined the next three years of my life at the NIH. Ted was a good scientist, and he taught me about gasses and diffusion of gasses. It was all about putting oxygen into the body through an artificial lung, which also removed carbon dioxide. All about the impedance of the solid membrane, the importance of blood flow patterns. It set me on a research course for the rest of my career.

We first tested Ted's spiral coil oxygenator on dogs with arterio-venous bypasses, with the assistance of our veterinarian, Joe Pierce. When my children were young, and they had their own dog, I remember one dinner in Concord when they looked at me with wide, sad eyes when I told them about my experiments with Kolobow on dogs.

"Would the dogs die?" Liza said, holding the ears of Darwina, our Beagle, blocking the dog's hearing.

"Yes. Often. And if they didn't, we would often euthanize them so we could study their lungs."

"What's euthanize?"

"Kill."

Liza burst into tears, and Nikki rolled her eyes at my blunder, and that was the end of that dinner.

It was worse when Liza was a strong-minded 13-year-old, and she took me to task when she learned that a limited number of seals in my Antarctic experiments died. She looked at me with outrage, tears in her eyes. How could I tinker with seal's lives? The problem extended beyond the home. I remember when I was stopped from going to my lab by Greenpeace activists at MGH holding picket signs with my name on them.

Their concerns and Liza's concerns were understandable, but I didn't have much patience for them. I held my sights on the goal of saving human lives. Animal experiments are a necessary step in medical innovation. They are a part of the regulatory process for good reason. We cannot ask humans to undergo trials of new drugs or devices before we have sufficient evidence that these treatments are safe and effective in animals. In my lab, we learn enormous amounts from working with mice, on sheep, pigs. However, there is no reason for these experiments to be painful to the animals, inhumane in any way. I served on animal care committees for years, shutting down labs where the researchers treated the animals poorly or had unsanitary conditions. I advised on animal care protocols, and hired deeply ethical, thoughtful animal care specialists. I always thought it was important to care for these animals well. There often are ways of tinkering with the experiments to come up with a less invasive, less painful route to the information. We must carefully design animal experiments. They are crucial to scientific and medical advancement.

We soon switched to lambs though, which were on the whole less excitable creatures. Lambs also have lungs that are anatomically more comparable to humans than those of dogs. The lambs allowed us to test the oxygenators for hours or days at a time. Some six months later we set our sights on testing the oxygenator on fetal sheep. Working on fetal sheep was more delicate than working with natural-born lambs, of course, and even more daunting. The sheep weighed about three kilograms. The plan was for them to have blood taken out and put back into their bodies after passing over small (0.25 meters squared) membrane lungs like those Ted had constructed. This would be the first test of a device that could exchange oxygen and CO_2, ever. If it worked, it would have the potential to treat millions of sick people in the intensive care unit.

Our lab team prepared for these oxygenator trials by reading up on the subject. I remember pouring over Geoffrey Dawes book *Fetal Physiology*. Dawes was a renowned physiologist at Oxford University who had pioneered the study of near-term intrauterine fetal sheep—for sheep, term is day 148 of pregnancy. Reading Dawes' excellent work helped prepare me for what lay ahead. Even so, this was tricky and finicky and I was nervous to carry it out.

In the midst of that period, Larry Aronson, a biochemist who worked with me in Building 10 at the NIH, figured that I needed to get out of my head a bit—and he was right. He and his wonderful wife Joan took me blue crabbing in Chesapeake Bay, a pastime that did the trick, and that I repeated several times with Nikki, before and after we were married. All minds need rest.

Even once the trials were in place, Ted continued to innovate. He came up with another excellent idea for how to cannulate a fetal umbilical artery and a vein. He dreamed up a stainless-steel spring reinforced lycra catheter, and wanted to build it in our lab machine shop. One of the difficulties we had to overcome was the simple question of sourcing suitable materials. We went to DuPont for the lycra polymer we needed, but to our amazement and disappointment, they refused to give us what we needed. As it turned out, they were worried about the potential for lawsuits if their materials were put to use in a medical apparatus.

I was baffled and deflated, but Ted refused to be stymied by the conservative reflexes of industry. One day he arrived at the lab with a broad smile stretched across his face. I couldn't understand what made him so happy—we were up to our ears with challenges making our invention. As I watched, Ted pulled a lycra brassiere out of his bag! I burst out laughing. I knew straight away what he had in mind, although I never would have thought of it myself. He wanted to extract the lycra from the bra! And that's exactly what he did. He dissolved the bra in dimethyl sulfoxide, an organic solvent, in order to obtain the remaining lycra in a polymeric form he could work with. If you can't work with industry, work around them. I learned so much from that man about the importance of tenacity and imagination in good research work.

Thanks to Ted, we now had everything we needed: flexible catheters, a pump and a membrane lung system. And so we began our work with our incredible team and new instruments. We carefully gave a C-section to a pregnant sheep, and took the baby out. Next, we cut the umbilical cord and took that stainless steel spring-reinforced, bra-derived lycra catheter, and stuck it into the vessels in its umbilical cord, hooked that up to our extracorporeal membrane machine and transferred the lamb to an aquarium. How was the lamb doing? Was it breathing? Jerry Vurek, an engineer in the lab, designed and built an in-line blood oximeter which allowed us to measure fetal oxygen levels with accuracy. And so we began perfusing these heparinized, near-term isolated lamb fetuses with artificially oxygenated blood. With a spirometer we measured oxygen consumption of the near-term isolated fetal lamb at 8 milliliters of O_2 per kilogram of weight, each minute. By satisfying this need through artificial mechanisms, miraculously, the premature lamb stayed alive in our aquarium! We had re-created some of the conditions of the womb, successfully delivering oxygen to the little baby lamb! And so we continued, and we perfused these lambs with our "artificial placenta" for up to 72 hours—a big step forward in our research.

Altogether, it was an amazing accomplishment, and I was thrilled to have experienced it from start to finish, always in attendance, assisting

Ted as best I could. The NIH made a short film of us performing fetal perfusion. The press caught on, imagining that women could grow babies in aquaria. Their imaginations went wild! Moms wouldn't need to carry babies in their bellies anymore! They could take them out whenever they wanted to! But of course, our solution for lambs would only last for a few days, as explained in *The New York Times*'s coverage of the experiment: "An Artificial Lung Supports Life in Unborn Lambs up to 2 ½ Days," an article by Jane Brody published in May of 1969. I was quoted throughout, my first time in the popular press. Even more importantly, *Science* published our paper.

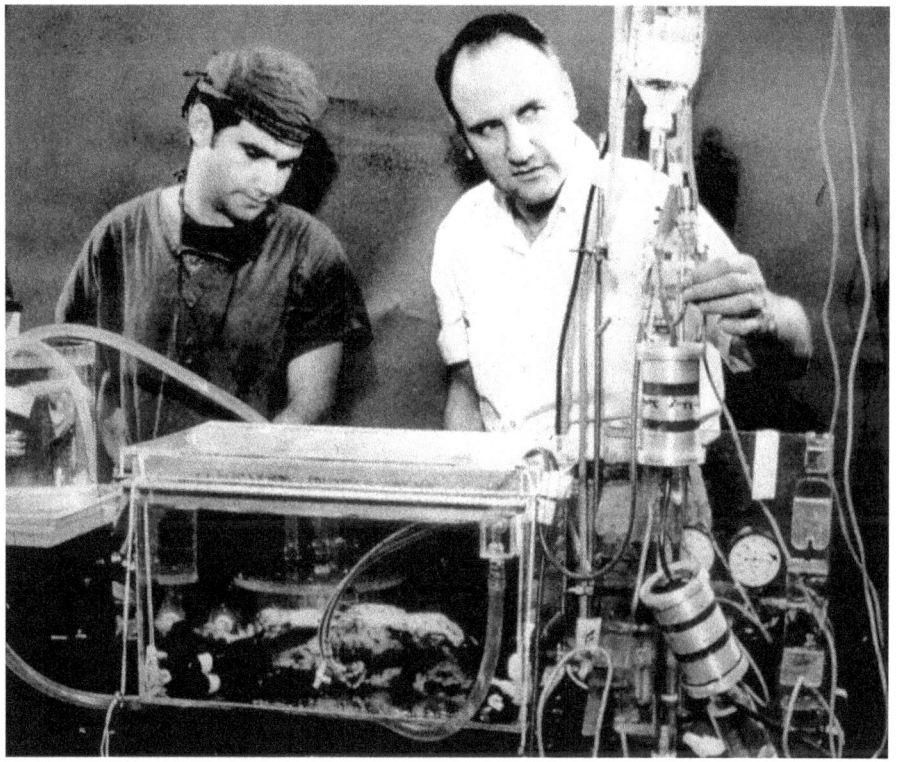

Demonstration of the first artificial placenta with Ted Kolobow in 1969.

* * *

Ted basked in all of this like a proud father. We all knew, though, that there was more work to be done to understand how to save human lives. Now that the concept had been proven in sheep, Ted said it was time to study newborns. Humans. My first time working on a study with people. My mixed feelings about working with lambs, naturally, multiplied in all directions. I was nervous, excited, hopeful. These babies were sick—they were oxygen-deprived (hypoxic) and needed help or they would die. With our new instrument, we hoped we could perfuse them, oxygenate them and save their lives. We went through regulatory approvals, of course, and we traveled to the University of Puerto Rico hospital in Hato Rey, with high hopes.

We tried. We worked with five babies. However, our first five attempts at using ECMO failed. All the babies responded poorly to anticoagulation, bled into their brains, and they died. Cerebral hemorrhage. It was very disappointing and depressing. I wondered, should we keep pursuing this approach to improving oxygenation?

The research community said yes. What I learned in the wake of Hato Rey is that failure is an inevitable step toward innovation. Urged by doctors in several hospitals to try again, we studied hypoxic newborns at Children's National Hospital in Washington, D.C. with a neonatologist, Gordon Avery. Unfortunately, the results were the same. We realized preemies can't be anticoagulated, that the requirement to anticoagulate with high levels of heparin (at five milligrams per kilogram of body weight) caused hemorrhaging. This hasn't been solved: Even to this day, premature babies are more challenging than full term infants, and are not put on ECMO. We decided that it was too risky to continue the ECMO studies on babies.

With that new insight, we proceeded to make necessary adaptations to Ted's invention, scaling it up tenfold in membrane area, allowing us to treat adult patients with acute respiratory failure at NIH. Finally, our first patient there, a young woman, survived a three-day veno-arterial ECMO procedure. I can still perfectly remember the daunting hours, waiting for the results of the procedure, and the elation we both felt when she stabilized and recovered. The weight of our failures in Puerto Rico lightened. Our persistence had finally paid off. What we all had worked on together held real promise for saving many lives.

Samuel Beckett, the great Irish writer and playwright, described this creative process best: Try again. Fail again. Fail better. This rings true for all human endeavors. This is what I learned at the NIH: a combination of learning, adjusting, advancing. Sometimes the scientific effort won't flower in your lifetime, or in your medical lifetime, or in your limited work time. Some of the technologies need to be developed, and they're not really there yet. But they will be developed, because they're useful for humanity.

My time at the NIH was almost all-consuming, and very exciting for me as a junior scientist. I couldn't imagine a better mentor than Ted, and I often thought of him later, when I taught others.

It was 1970. The Vietnam War was the focus of all Americans. American men were dying, more than a hundred a week. Nixon's approval ratings were still high. Commander Charles Brodine of the U.S. Navy called Ted's lab. "Hey, we have this thing called 'Da Nang lung,' soldiers with terribly injured lungs from blasts. Can you do ECMO in Da Nang?" He explained: The task was to perfuse soldiers with severe traumatic lung injury resulting in ARDS. We told him we'd discuss it.

When I raised the matter with Ted, he looked at me silently and nodded. No words were needed. I understood that he was in favor of sending me straight into the Vietnam War with ECMO. I was thrilled and terrified. How far would I go, and how much would I risk, for medical innovation? This wasn't an easy decision. I was newly married to Nikki, and leaving my peaceful life to go to a theater of war sounded awful. But again, here was a chance to save lives by applying our research. I was assured that I would not have to fight, nor would I be deployed too near the front lines. So I told Ted and Commander Brodine that I would go. Because it was wartime, they explained to me that we could get plenty of funding. The U.S. Navy and the Department of Defense funded the undertaking, and this time we received critical help from industry: Litton Systems fabricated the membrane lungs.

Gathering up just a few belongings, I boarded a plane and went from D.C. to Los Angeles, L.A. to Honolulu, Honolulu to Taipei, finally arriving at the U.S. Navy Hospital at Da Nang, where I was to spend March and April

1970. I met my team: Tommy Wonders, the Midwestern Navy corpsman with a wicked sense of humor (who came with me later to Antarctica); Jack Ratcliff, a terrific surgeon; and an anesthetist who went by the unfortunate moniker of "Killer" Kopriva. How did he get that name? Sorry, I never asked.

What I did ask was for sheep on which to practice ECMO. They were imported from Thailand just for us, and Tommy Wonders tended them. After they arrived, the sheep were spray-painted blue. I'm still not sure why. Maybe it was Tommy's crazy sense of humor. Maybe the color had some real function—maybe he thought they would turn pink with ECMO. But Jack, Killer, and I practiced ECMO perfusion on these blue sheep. The membranes from the Litton Corporation were huge: two and a half square meters of surface area. We discovered that the membranes all leaked, and the technology wasn't excellent. But this was a key moment in driving our experimentation forward, at least on blue sheep.

Even so, I was totally unprepared for what I found at the hospital in Da Nang. The conditions of war were a shock to my system in every way. Severely injured people, both civilians and military personnel, would be brought in. I'd

Vietnam, 1969.

pitch in and we'd operate all night on them to try and save their lives. It was awful. An intensive care unit where nobody was older than eighteen or nineteen years old was a tragedy. I was used to visiting an ICU where all the patients were seventy or eighty, ninety years old. I remember African Americans from the South and Iowa farm boys missing arms and legs. They had been blown up by various things, mainly tripped landmines. If they got really sick or septic, but could survive the trip, we put them on an airplane and sent them by medevac plane, to Yokosuka, Japan to get better care. Others couldn't even get that far.

It was a grim fact that many triple amputees committed suicide when they returned to the U.S. I remember discussing with the other doctors what we might do in such devastating circumstances, and I felt hopeless. We knew that we were unlikely as doctors to be out walking anywhere near where landmines laid in wait. But we were also well aware that nobody there was truly safe. I had always been anti-war, what the soldiers called a "pinko liberal," as were most of the other doctors I worked with, but we didn't say anything. Our charge was to take care of these poor, injured boys, most of whom weren't pinko liberals but had gone to war at their country's call.

I tried not to think about the senselessness of war. I tried to focus. And still, the working conditions made it difficult. At NIH, I had access to every resource or piece of equipment I might need. But the spartan nature of the hospital, combined with the climate, combined with the patients, shocked me. And it became increasingly clear that this wasn't the right site for ECMO experimentation or innovation: this was a mess. But I was surrounded by good doctors, especially very good anesthesiologists, and I tried to focus on learning from them. Killer Kopriva told me about the excellent anesthesia residency at MGH, which he had just finished, and my interest was piqued.

Vietnam moved me along in my mind, but also kept me at loose ends. The sleep deprivation was terrible. On an almost nightly basis, I would be woken by the sound of air-raid sirens going off. I would stagger out of my little air-conditioned bunk in pajamas and shuffle to a tunnel that had been dug beneath our Quonset hut for this very purpose. An effective form of psychological warfare by the Vietcong (VC), ensuring that we rarely got a good night's sleep, aided by the rats and mice that infested our shelter-tunnels. If

I ever felt myself nodding off to sleep, the sound of those rodents scurrying along the tunnel floor was more than enough to keep me awake until the bombing subsided. Then, I would stagger back to bed, in the "crap of dawn" to snatch what was left of the night's sleep before returning to work.

This was a war with no holds barred. One night, I was in the operating room, and the VC blasted us. Clearly breaking the rules of war, they sent a rocket at the hospital, disregarding the large red cross painted on the roof. It was loud and chaotic. I was terrified. I could hear the helicopters travel up into the air as soon as the rocketing began—they didn't want to get hit. I remember the sound, and the terror. I honestly don't remember much else, or I just can't. Later, we learned that the Vietnamese cleaning women had been creating maps of our base and shared them with the VC. But why did they aim at the operating room?

To get some much-needed sleep, I traveled north of Da Nang on the hospital ship Sanctuary, to I Corps, what we called the area the furthest north before hitting North Vietnam, where fighting was most extreme. These hospital ships would stay two to three miles offshore to avoid being caught under fire or combat. There, in the quiet and moments of rest, I felt some homesickness. I found the radio room on the boat, made friends with the operator, and called home, and spoke to Nikki. My passion for radio always came in handy, especially when I was far from home.

I never got the opportunity to treat severe ARDS, or to use ECMO on the soldiers, the purpose of my trip to Vietnam. Of course, on the ground, nobody showed concern about why or how I had come to the theater: individual intentions don't count for a whole lot in war. And besides, despite the ongoing carnage, this war was at last winding down. Most folks, like me, were dreaming about getting home at that stage.

As terrifying and clarifying as my experience of the Vietnam War was, it was brief. Despite the rocket bombardments, I was rarely in real danger. I was able to spend my weekends treating Vietnamese civilians, which gave a sense of purpose to what might have felt at times like a wasted trip. However, more than once, I hung up an IV on somebody and walked away. When I went back, the IV was gone, disappeared. Stolen. These Vietnamese were so poor and short on resources, I think they would steal the IV for their relatives. I learned to watch the IV, or to ask the nurses to. It was damn hard to do good medicine.

One of my worst memories was of a helicopter medevac off the Sanctuary ship to the hospital in Da Nang. A young sailor would get drugs somewhere then shoot up on the ship. Like so many, he just wanted to escape the war by any means available. But then they'd stop breathing and be blue. The captain would call me. Since he was blue, they were thinking it was a lung injury, and that ECMO could help. But with lung X-rays, I could see it wasn't anything for ECMO. When the kid had passed out, they threw up, breathed it in: it was aspiration pneumonia.

I sat in the helicopter, which of course had no windows—the VC routinely shot them out—so they are incredibly noisy. I couldn't hear his heart. I tried to listen to his breathing, but I couldn't. It was so loud. How do you know how the patient is doing? Is he still alive? All I could do was feel the pulse. It was terrifying. How do you take care of somebody in a noise machine? You looked to see if they were pink and had a pulse. I was in a vibrating noise machine, vibrating like crazy from the motor and the two props slicing through the air overhead, and the breeze flying through. There was no way of knowing if he was still alive until we landed at base; every second of the journey hung heavy with fear. I was relieved when we got him to the hospital, and we nursed him back to health.

There was not much room for medical innovation in the Vietnam I experienced. There were no patients with blast injuries to their lungs when I was there. Perhaps it was a detour, not a failed experiment, but certainly the wrong conditions for one. Even so, I learned so much from those experiences. Fundamentally, I learned how damn lucky I was that I wasn't a soldier there, that I was able to get away to safety. I continued to learn the importance of quiet and rest, and the necessary conditions for thought. As an innovator, I continued to learn that you don't always win. And I learned that I was interested in anesthesia, in intensive care, and in caring for critically ill patients. I was scared, but I learned that I still did well under pressure. I was just eager to get out from the pressure of war.

Ted Kolobow modeled innovation and experimentation. Vietnam taught me how to persist in inhospitable circumstances. Both were preparation for what came next: building my own lab at MGH, bringing that sense of risk and experimentation back home. But I never stopped thinking about the importance of testing and tinkering in the field, of continuing to innovate under extreme conditions, on the ice, across the iron curtain, in the ocean depths.

CHAPTER 9
The Dome

* * *

Study of me as Morton, by Warren Prosperi.

YOU CAN BE SURE THAT bearing a likeness to William Thomas Green Morton was not one of the qualifications initially sought by the august Harvard committee that appointed me in 1993 as the third Anesthetist-in-Chief at Massachusetts General Hospital (MGH). Morton, of course, was a dentist and physician who conducted the first public demonstration of the use of inhaled ether back in 1846. This marked him as a modern pioneer of anesthesia. His importance was unquestionable. What did it matter who looked like him?

Yet, when it came time to select a model to pose as Morton for an MGH-commissioned portrait that would hang in the Ether Dome—the majestic, sky-lit, copper-domed structure designed in the late 19th century by Charles Bulfinch for the purpose of isolating the screams of anguish of patients having to undergo surgery in pre-anesthesia times—I was the top choice. It was not all that far-fetched a choice. If you squinted you might agree I had a physical likeness to Morton—my dark, thick hair, prominent nose, the shape of my face, my caterpillar-like eyebrows that could, at times, impinge on my sight. And then there was the fact that the hospital decision makers decided to search first among its senior staff to find suitable models—due mostly to their fields of clinical expertise and, as closely as possible, their appearance—for the historic figures.

Even so, I did not jump at the chance to take the part. I was fully absorbed with the non-stop demands of the Chief position, and concerned about the time it would take for me to be a reliable Morton for the tableau. I would need to commit to makeup and costuming sessions before we reenacted the historic moment.

After learning of my selection, I came home and told Nikki about my day. The Morton news slipped in almost as an afterthought. She stopped me right there.

"You as Morton? You have to say yes. Think about it."

I had plenty of other things to think about. But she persisted.

"Warren! That painting will hang in the Ether Dome of the Bulfinch Building long after you and I are gone. Whoever knows that you stood in for Morton, certainly our children and grandchildren, will be able to tell the story of not only Morton, as you love to do, but also about you."

I shrugged. If Nikki felt so strongly about it, I would make the time. In retrospect, I am very glad she prevailed. I had a great deal of fun becoming Morton in authentic costume and mutton chops (they kept sliding off my sweaty face) for the fantastic painting by Lucia and Warren Prosperi. In fact, I approached it as an acting challenge, drawing on theater skills I had learned performing in musicals at Camp Zapol and at MIT. The "Ether Day" painting now hangs on the front wall facing a raked amphitheater within the Ether Dome.

Morton's real story was a real cliff-hanger. I have told it many times in live presentations for visitors at MGH, for large events (some held by MGH during its annual celebration of Ether Day), and for the media (an internet search will turn up a few). Two books were devoted to it: *Tarnished Idol* by Richard J. Wolfe and *Ether Day* by Julie Fenster. Here is a condensed version of the story.

Morton, a 27-year-old Beacon Hill dentist revolutionized surgery when he demonstrated that by inhaling ether a patient could become insensible to pain. Morton experimented by inhaling sulfuric ether himself and giving it to his pet dog. He then administered it to patients during tooth extractions. Word of his success rolled down the hill to MGH, where he was invited to demonstrate the power of his new potion. On April 16, 1848, a skeptical MGH medical staff gathered in the hospital's operating theater (now, the Ether Dome). Morton rushed in a few minutes late with his brand-new, just-completed contraption, an inhaler, hand-blown to his specifications by a glass blower in Boston. Through the mouthpiece of the device, the patient inhaled a vapor emitted from a saturated sponge inside the inhaler. The patient, Gilbert Abbott, remained almost completely quiet during surgery on a tumor on his neck. When finished, he professed having felt no pain. The surgeon, Dr. John Collins Warren, looked up at the mesmerized gallery of physicians and is reputed to have declared: "Gentlemen, this is no humbug!" Morton called the special ingredient "letheon" in an effort to keep it proprietary. In fact, it was ether, and Morton's claim to have discovered ether as an anesthetic caused controversy, distress, and tragedy. Another claimant, Harvard Professor Charles Jackson, an analytical chemist whose help Morton had sought while searching for the right chemical formulation, is said to have become crazed by the dispute; years later he committed suicide. Twenty years after the historic MGH demonstration, Morton died on a hot July day in New York City. He was there while continuing to press his case against Jackson in hopes of obtaining compensation for the government's use of ether during the Civil War. Nonetheless, Morton's place in history as the first person to publicly demonstrate—and thus bring to world attention—a means to perform surgery without pain is, so far as I know, undisputed.

I confess to a feeling of pride and perhaps even immortality about being there in that painting in the Ether Dome. This building, with its historic roots, symbolizes the institution which has been my base of operations for all my professional life.

By the time I played Morton, I had been at MGH for two decades. Dick Kitz convinced me to join MGH as a first-year anesthesia resident in 1972. A wiry midwesterner from Oshkosh, Wisconsin, Dick was just three years into his tenure as the second chair of anesthesia at MGH. He gazed at me across a large walnut desk with his gray-blue eyes and impish smile. My strong record at NIH and my desire to continue both research and anesthesia were on the line. Dick was surprisingly quick with an offer. "Warren," he said, "if you come here, I will support you in getting an NIH grant to continue your research on ECMO and will give you the time from clinical responsibility that you need."

I could not have asked for more. I accepted. Dick followed through. With his help, I received an NIH RO1 grant, the oldest and most prestigious grant awarded by the National Institutes of Health for independent research and the first ever awarded to a medical resident at MGH. His support, along with the support of others at MGH, enabled me to hone my clinical skills while continuing my research on artificial lungs. Without that encouragement, I could never have succeeded in becoming a physician-scientist, nor, I would venture, would I have followed Dr. Kitz as Anesthetist-in-Chief at MGH.

Once Dick got me installed at MGH, I stayed and I thrived. I relished coming to the hospital, not only for the fascinating work but also for some very special people. A few were giants with shoulders I would stand on. Others just made the work really fun! To do justice to them all would require another book, here I will mention just a few.

Edward Lowenstein occupied the lab next to my first lab at MGH, and took me under his wing. We quickly became like brothers. I learned over time that he was a survivor of World War II thanks to the kindertransport that took him from Germany to England. Ed was reunited in Cincinnati with his family and became the first cardiac anesthetist in our hospital, pioneering studies of injectable anesthetics. Ed loved research and medicine, and had a healthy sense of humor about the high-handed and otherwise infuriating treatment anesthesiologists received at the hand of some MGH surgeons. His positive attitude and mentorship carried me through my early years at MGH. I would come into his office, find him half-hidden behind chaotic stacks of manuscripts and books. We talked about our research, clinical challenges, and personal lives. We supported each other through daily incidents in the OR.

We even helped each other with our growing families. "Mike had a run-in with the law," Ed said one day about his teenage son, looking for help from Nikki, who was a lawyer. I counseled him not to worry too much. Some time later, I pulled Ed aside in the hallway. "David's school called. It seems David said he turned in his sixth grade homework, when he hadn't." Ed counseled me to be patient. I also sought his counsel when David somehow set our neighbor's lawn on fire. Ed calmed me down pointing out that I had experimented with explosions in my neighborhood as a kid. Forty years on, it turns

With Ed Lowenstein.

out that Ed and I were right to calm each other. Mike is now a corporate lawyer, and David… Well you've read by now about our work together in Antarctica and building Third Pole, so I guess it all turned out okay, though he still enjoys a good fire.

Ed eventually left MGH for a stint leading the Department of Anesthesia at Beth Israel Hospital in Boston, and then returned to MGH in a senior position after I became Chair, and is now retired. Our families continue to be close, down to our grandchildren.

My positions at MGH and Harvard over the following years progressed through Assistant, Associate, and in 1985, Professor of Anaesthesia. In 1992 I was given the enormous honor of being appointed the Reginald Jenney Professor of Anaesthesia. Endowed by the founder of the Jenney Oil company, this position gave me the security of remaining a full professor at Harvard until I either relinquished the title, or died. My own appointment had only become possible when Henning Pontoppidan, the first incumbent of the Jenny Professorship, had decided to step down. Henning was another mentor and true friend, not to mention an icon, having established the first respiratory intensive care unit in the U.S. at MGH in 1961. Henning deserves volumes. You can read more about him in the collection of essays *This is No Humbug! Reminiscences from the Department of Anesthesia at the Massachusetts General Hospital* (2013).

In academic medicine, every researcher must secure their own funding for research: As John Bunyan said in *Pilgrim's Progress,* "every tub on its own bottom." The largest source of that support has always been the National Institutes of Health. Writing grants to receive NIH money is an all-consuming task, and the success rate is low: For grants from the National Heart Blood and Lung Institute, the success rate hovered around 20 percent. Some departments may have funds to assist, but none have enough to fund all the promising research proposed. I had a lucky stretch in receiving coveted NIH funding, but my luck was running thin in the early 1990s. At that point, I considered whether to persist in putting my efforts into grant writing, or pursue another option—becoming a chair of an anesthesia department. Chairs of academic medical departments are prestigious and generally well-compensated positions. At least until the mid 1990s, they were also very powerful in steering the direction of their departments. I reasoned that as a Chair, I could prioritize research excellence, encourage and recruit promising researchers while maintaining the level of clinical services required of the anesthesia department. If I was lucky, I might even be able to keep my own research going—I chose the Chair option.

In 1993, in fact, I was offered the Chairs of anesthesia at MGH and at Johns Hopkins Medical School. Both were strong institutions, but it was not a close call. My strong ties to Boston and MGH, and the critical, unswerving support from Ed Lowenstein who was on the Harvard appointments committee, won the day. Another factor also influenced my decision: I knew there was a possibility that the MGH license to the patent on inhaled nitric oxide had a reasonable chance of bringing in a royalty stream, and if so, my laboratory would receive a share of those funds. I began my tenure as Anesthetist-in-Chief on April 1, 1994—April Fool's Day. It was no joke.

Before I was selected as Chief, administrative concerns had not been central to my clinical work nor research. In fact, I avoided involvement in administrative details as much as possible. My leadership experience had been with relatively small teams, whether in Antarctica or in my own lab. I had not wanted to involve myself in the workings of a bureaucracy. Now, however, large-scale administration became a critical part of my daily responsibilities. I had a lot to learn, quickly. I enrolled in an outstanding Harvard Business School course for leaders in health care, learning about organizational behavior and budgets. That was enormously helpful, eye-opening, and tough.

Easier for me was the opportunity to formulate, convey, and pursue my vision for the Department. I invested in the burgeoning area of molecular biology and its relevance to the neurobiology of pain, an area of utmost importance to anesthesiologists. I pledged to focus on improving the residency, particularly with respect to requiring more teaching of trainees during surgical procedures. In the clinical arena, I acknowledged what then felt like an urgent need to meet the challenges of managed care, bringing with it cost constraints and a shortage of qualified anesthesiologists. I also oversaw a major structural change when Partners HealthCare System was formed as a collaborative effort of Brigham and Women's Hospital, Mass General, and a number of smaller area hospitals.

My efforts were not always met with enthusiasm, and they were sometimes met with resistance, but I always managed to build support and forge forward. I am gratified by the advances made by the department in all areas, despite serious headwinds in some, during my fourteen-year tenure. In research, we established three Harvard tenured chairs in anesthesia (the Kitz, Beecher and Zapol chairs), and shored up funding for two others (the Mallinckrodt and Dorr). We recruited a leader in the neurobiology of pain, Dr. Clifford Woolf, who published extraordinary findings consistently each year and fostered a cadre of distinguished scientists in many related areas, securing precious space and funding for them.

In education, I supported establishing a permanent simulation program and center under the leadership of Jeffrey Cooper, which has become a model for such programs around the world. At the dawn of the introduction of electronic technology into the medical arena, we assembled a small group of savvy individuals who developed a digitally-based teaching program for anesthesia. In the clinic, we designed the first MGH department-wide electronic network named, appropriately, "Etherdome." That was not instantly recognized as a benefit. In 1997, when I asked the head of the hospital for funds to purchase computers to build the network, he raised his eyebrows and asked, "What do you need computers for?" It was an uphill battle. In clinical care, we built a robust capacity to administer regional anesthesia (that is, blocking the nerves in, for example, your shoulder only, rather than putting you into a general state of anesthesia). We staffed up to deliver anesthesia for the new obstetrics service at MGH and also became one of the first major academic hospitals in the U.S. to offer acupuncture.

Other demands of the job were significantly less elevated, though no less important. Early in my tenure as chief, a late-night fistfight broke out in the OR between a cardiac surgeon and the attending anesthesiologist over a plate of fried onion rings explicitly (and infuriatingly, I suppose) reserved for anesthesiologists. I called both of them into my office, along with a deft and gentle MGH psychiatrist. The two combatants needed a refresher on the skills they should have learned in kindergarten. After some very uncomfortable moments, they were listening to each other and looking each other in the eye. When they parted I had some hope that they would not be coming to blows in the hospital again, and maybe others would take note.

Sometimes, the personal and the significant intersected. One day I answered an early-morning call at home. "Dr. Zapol," I was told, "One of the staff has overdosed on fentanyl. We found him unconscious in the on-call staff room." I jumped out of bed, turning to Nikki, who was now fully awake. "There's been a drug overdose," I told her, already getting dressed. "This is the worst day of my life." I rushed to the hospital to tend to the matter. I cannot share the details of this story for privacy reasons. But the lesson I learned that day convinced me that I had to do something to prevent such incidents, and worse, from happening on my watch. Fentanyl, a powerful opioid pain medicine, is responsible for a large percentage of the tens of thousands of deaths worldwide resulting from illegal opioid use. It is far more effective than morphine and heroin in reducing pain. Addictive, it can cause relaxation and euphoria. Even in very small doses, it can interfere with the brain's signals to the respiratory system and result in death. Anesthetists are by necessity given access to a whole range of potent anesthetic drugs. They live highly stressful professional lives, sometimes characterized as "hours of boredom and moments of terror." That they would break was not surprising.

This particular staff member survived that day, and with help continued on to a brilliant career. But it was not to be the only drug incident among the residents and staff during my tenure. Luckily, none ended in death. I stayed calm and carried on with a focused effort to prevent any further overdoses among the residents and staff. Battling considerable opposition ("Nobody will apply to our program if we infringe on their privacy," I was warned), I instituted a randomized drug-testing program in the department, and tasked an interested staff member with working on a research study that could provide proof of its efficacy. Other programs around the country took note, and some started their own programs. I became a Cassandra of the inevitable harm that opioids could cause not only in hospitals, but on every street corner in America.

The relationship between opioids and anesthetics was even more complicated. Prior to the overdose incidents, I was part of an effort to secure corporate support for the hospital from one of the largest opioid companies, Purdue Pharma. MGH had an enviable track record obtaining large corporate support for research, including from Mallinckrodt, Hoechst Pharmaceuticals, Shiseido, and Bristol-Myers Squibb. Nikki was working in the MGH office that negotiated those agreements. When I was asked to join a hospital team to solicit funding from Purdue Pharma, I agreed. Our group journeyed to Purdue headquarters in Stamford, Connecticut and got an offer from the company for a substantial sum of money. There was, as I recall, one specific request: that a pain unit at MGH display a plaque crediting Purdue with its support. For a few brief days, it looked like we would be heroes. But when the Board of Trustees of MGH met to consider the transaction, they turned it down. While a legal document could and would draw a red line between that arrangement and any hospital decision-making, including purchasing, research, education and patient care, the Trustees decided the optics were terrible. Trust in the hospital was and will always be its most important currency. And, in case that wasn't immediately apparent, the *Boston Globe* ran a sensational article on this potential gift, highlighting the dangers of opioid use. This experience was high in my mind when I was confronted with the devastating effects of opioids in my department. The wisdom and prudence of the Trustees was all the more apparent to me when I decided to put guardrails around my staff.

The thicket of managed care did cost me some sleepless nights. Billing and money management assumed top priority. One day, Law enforcement officers appeared in my waiting area, demanding explanations for infractions in billing. "Not only am I running a kindergarten," I said to Nikki that night, "But you may soon find me wearing a striped suit."

And then there were the trends of medicine, and the way that we were buffeted by those winds. In the mid-90s, there was a prediction that insurance coverage of surgeries would be limited, leading to a decline in various procedures and therefore in the corresponding anesthesia. The press took to it immediately. In March of 1995, *The Wall Street Journal* ran a front-page article titled "Numb and Number? Once a Hot Specialty, Anesthesiology Cools as Insurers Scale Back." True or not—and it turned out to be not, mostly—it resulted in a dramatic decrease in applicants to our residency, as hospital managers decided that there was a national oversupply of anesthesiologists, and there was no need for training more. We had thirty slots to fill each year. I will never forget when Greg Koski, then chief of resident recruitment looking to fill those slots, called me one March. "Warren," he said, "You aren't going to like this. Better sit down."

"I'm sitting down." I was at my desk, looking out on the snowy courtyard in front of the Ether Dome.

"We matched eight." We were both silent. That was a shortfall of 22. I was dumbfounded.

Students had stopped applying to anesthesia residency programs. Some places were hit harder than others. Not a single U.S. medical student matched to any anesthesia residency program in the state of Pennsylvania that year. The next two years were similar. Ultimately, as the supply of U.S. trained anesthesiologists dipped, the price that hospitals were willing to pay for their talents went up, and we began to see a rebound. Supply and demand balanced out.

It took me a while to realize that this market aberration presented an opportunity. We were set up to train anesthesiologists, and there were plenty of people from other countries who wanted to come work with us. If U.S. medical schools were only going to deliver us eight residents, we could accept a host of well trained, intelligent, hard-working physicians from mainland China, Japan, Italy and Brazil, many of whom were finishing U.S. research fellowships or PhD training and looking for opportunities. Because we were at MGH, we were able to recruit from this foreign talent pool a group of brilliant teachers, clinicians and researchers. I am deeply grateful to all of them. Some became as close as family to us. One, Tong-Yan Chen, was a research fellow in my lab while we were working on the discovery of inhaled nitric oxide. A remarkable physician-scientist, he had left China for the U.S. after the uprising and massacre in Tiananmen Square. I asked him about that when I interviewed him. "I rushed to the scene as a medic," he said. "I had to help. It was absolutely horrifying, the government shooting innocent young people, like animals. I think I recognized some of them. I did what I could, but it wasn't enough. I was shattered." He paused and looked at me, "I could no longer remain in that country. That is why I came here to the U.S., with my wife."

I admired him greatly for his bravery, determination, intelligence and character, and had the enormous good fortune to have him and others as colleagues. Our MGH program became a magnet for foreign applicants. They knew to come to us. Not long ago, one told me of an old Chinese proverb:

"You don't need to see the wind. You can tell which way it is blowing from the movement of the branches on the trees."

∗   ∗   ∗

Of the outstanding researchers who I retained, recruited, and supported, I have a special place in my heart for Kenneth D. Bloch. Ken first came to my attention thanks to Nikki. Around 1993, Ken called her in the MGH tech transfer office for advice on how to proceed with protection of a discovery he had made related to endothelins. He was scheduled to present his findings at a prestigious Gordon conference and wanted to make sure he was covered. "My husband is working in that area also," she said. "Do you know him?"

"No," Ken replied. "But I know *of* him."

"You should call him."

"Really? He's much more senior than I am."

"You should call him!" she repeated.

Partners in Science. With Missy Flynn and Ken Bloch, at Liza's wedding in 2009.

He did. In no time we both recognized how much we had in common in science and in personality. He, like I, was laser focused on his science, and loved it. As Chair, I had established the Anesthesia Center for Critical Care Research with generous space secured on the fifth floor of the Thier building. Ken agreed to move there from his senior position in the MGH Cardiac Research Center in Charlestown. For the next two decades, we conducted a diverse and energized orchestra of junior and senior researchers in pursuit of clinically relevant advances in the general area of nitric oxide. Ken died at the age of 58 of pancreatic cancer. A terrible loss for me, for medicine, for the world. His legacy is evident in those he inspired and trained and in the work in the laboratory that continues to this day. Thank you, Ken.

I've spent much of this memoir discussing research, in one form or another. Research, though, is a by-product of relationships with others, of mentoring, of building teams to push scientific frontiers. At the time I was seeking the chairmanship, I had already nurtured and mentored many wonderful clinician-researchers and teams whose impact would be felt worldwide. As Chair, I knew I could also exert significant influence over the future of research conducted in our department and beyond. I will go so far as to say that this was what galvanized me the most about taking the job. I also knew, however, that as with all academic teaching hospitals, MGH was and still is, under immense pressure to keep clinical funds flowing. I hope I succeeded on both counts.

Boston is very cold in winter. In fact, having experienced the Antarctic in winter, I agree with Admiral Byrd. After returning from flying the first plane over the South Pole in 1929, he proclaimed that the coldest place he'd ever been was the Harvard Bridge—the one that crosses from MIT to Boston—in February. But the winds were and still are blowing this way, and have always brought me back. Here, I have been able to stand on the shoulders of giants and influence the future of modern anesthesia and critical care. Pretty hard to beat.

Portrait by Warren Prosperi on display at MGH.

CHAPTER 10

Home

* * *

WELL BEFORE I HAD GRADUATED from MIT, my mother made it clear that she expected me to settle down. She figured a family would anchor me. I knew that I wanted to be free to go wherever I needed to go and accomplish whatever I thought worth accomplishing. That was my main goal, my main thought. If I had passing thoughts of marriage and family, it didn't occur to me that it would take me off track.

"Dad, did you have any crushes on anyone as a kid?" Liza asked me recently.

"Liza, I've never been clear what 'having a crush' means," I replied, betraying my poor vocabulary about romance. "Who is crushed by whom?"

She laughed. "Did you have any girlfriends?"

"The truth is, I can't remember any girlfriends until I was well into my college years."

There were reasons. As a kid, I was probably seen as a nerd who was short, obese, and young. Stuyvesant was all boys. MIT was essentially that. Even in college, my relationships were not serious by today's standards. I think the most exciting thing we ever did was kiss. In my junior and senior years, I dated a few women from Wellesley. I would pick them up on my roommate's single-cylinder BMW motorcycle and rumble to concerts at Symphony Hall in Boston. I loved music, I think all my dates were musicians. I told my parents I wanted to marry one of them. She was tall and blonde and smart. But she wasn't Jewish—none of them were. My parents threatened to disown me. I was furious, but I gave in.

So, after I met Nikki in Washington, D.C. in 1967 and we got serious enough to talk about getting married, it did not occur to me that "getting hitched" (which was the only phrase I could muster when I asked her father for her hand in marriage) would alter my trajectory. Any awareness of that was in a blind spot.

I'm getting ahead of myself, though. Before we got serious, we had to meet. How did that happen? Well, once I knew I was heading to NIH, I turned to my good friend and former camp co-counselor, Joe Silk. Joe had a knack for keeping track of young ladies.

"Joe, who do you know in Washington?" I asked.

In those days, we didn't have the internet. Instead, we had our little black books that we bought at the Harvard Coop bookstore in Harvard Square. Joe flipped to the back of his and found a few names in D.C., Vassar grads mostly, including Nikki. He said he had met Nikki in Woods Hole on Cape Cod, walking on the beach. She told him she was "busy" at the time, but she agreed to give him her phone number. I called.

A male voice answered, repeating the number I had just dialed: "315274… Hello?"

I hadn't expected a man and I wasn't sure what to say. I stammered my way through it. "Uhh… is uh… Nikki.. Nikki… Kaplan, there?"

"No. Sorry. She's not here." It seemed to be an office rather than a home, which gave me some relief. "You can try later." He hung up.

He didn't tell me I had reached the CIA in Langley, Virginia. Eventually I would find that out and a lot more about the person I was looking for.

I called back later. This time it was a female voice. "Hello, 315274."

"Hello, is this Nikki Kaplan?"

"Yes..?"

"Hello, Nikki. Remember Joe Silk, in Woods Hole?"

"Oh, Hi, Joe," Nikki replied, a lilt in her voice. "Yes, I remember you. Where are you?"

"No, Nikki. I'm not Joe, My name is Warren Zapol. I'm Joe's friend. He gave me your name and phone number. I just moved to D.C."

"I see." She paused. "So, what are you doing here, Warren?" She didn't sound terribly disappointed that I wasn't Joe.

"I'm a doctor, just arrived from Boston and beginning work at NIH. Would you like to meet?"

"Oh, that's interesting," she said. "NIH is supposed to be an excellent place. I have gotten to know a little about it because I have been looking into whether they can help my father."

"I hope it isn't anything serious." I offered.

"I think it might be," she replied.

"I'm sorry." I said.

She thanked me for my concern.

"So, can we meet?" I asked again.

She paused. "I would like to, Warren. I would really like to talk with you about NIH. I should tell you, though, that I am involved in a relationship. It may last, but it may not. I know this sounds odd, but if you want to call me back in say, a couple of weeks, I hope to be clearer about seeing other people."

We hung up on cordial terms. I didn't dwell on either the rejection or on the potential. I had plenty to do establishing myself in the lab at NIH, and I also decided to connect with other Vassarites on Joe's list. A few weeks later, I tried Nikki again. This time, she welcomed my call and we agreed to get together.

On our first date in the fall of 1967, I took her to the Officer's Club in the D.C. Navy Yard on a night when they had dinner and dancing. It was a pretty place with large windows overlooking the harbor. We did not lack for things to talk about, surprising each other with stories of our world travels, hers more extensive than mine, having lived in the Philippines as a child. When I asked her why she had lived there, she told me her father had been an undercover agent for the CIA: "Yes, that's where I work, too. It's the office you called to reach me—that was Langley!"

We were secretly gratified in discovering that each of us were Jewish. I think she was the first Jewish girl I had dated. I thought she was wonderful, gorgeous, just terrific. As an added bonus, my mother would be thrilled. Nikki was under no such pressure from her very secular Jewish parents, but being Jewish didn't count against me, either. I also surprised myself by taking to the dance floor, something I hadn't even done at my Bar Mitzvah. Dancing to the Beatles wasn't that hard—I could jiggle in my own orbit. Maybe I even tried to lead her in a more classic dance. I think she's never forgiven me for duping her into thinking this was one of my nascent talents that we would enjoy together if our relationship progressed. I had, and have, no such talent.

I asked her out again. The second time, I picked her up in my British racing green Triumph TR4 convertible. Its prior life in Boston had been hard. Among its defects was that the hood randomly popped open, a stunt it pulled while we were driving down Wisconsin Avenue. I came to a stop out of the stream of traffic, stuck my arm out and around the window to grab the hood. When that didn't work. I asked Nikki if she would get out of the car and close the hood because I had to remain behind the wheel. When she did, I didn't remove my hand fast enough, and the hood slammed down on my finger.

I was injured and bleeding, but determined not to show my pain. I calmly instructed her to find the right lever and release the hood. I wrapped my bleeding finger with rags from the back seat. It began to swell to double its size, but I was not deterred. We crossed the street to a Chinese restaurant. We ate with chopsticks. Nikki was much better with them than I was, but then again, I was handicapped. It wasn't the most auspicious date, just a stumbling step toward a lifetime together.

On a later date I invited her for dinner in my apartment on O Street in Georgetown; the building was next to a playground and across from a small, ornate gift shop. I cooked swordfish with hot sauce for dinner. She managed one bite. I silently agreed it was awful and softened the impact by playing some of my favorite classical music. I rarely was the dinner chef after that, but we attended symphony regularly.

We were with each other a lot during the following months. On weekends, if I wasn't engrossed with my work at NIH, I'd take Nikki on the motorcycle to go fishing for catfish in the Potomac. She was not as squeamish as I was about removing the hook, and she managed to ignore their unnerving barking when she prepared them for a fish stew dinner.

Along the way, we discussed the matter of her father. It had been the first order of business when I had called her, and it came back up in force once we started seeing each other. He was succumbing to a kidney cancer that had metastasized to his lung. I told her what I knew, which may not have been of much help, but at least served as comfort—a sympathetic ear. When we discussed getting married, the shadow of her father's illness fell across both of us. I felt it was not a good time for her to be making the momentous decision to marry me, but she felt an urgency to get married before he died. I knew she was a very special lady and I didn't want to lose her, so we moved forward with our plans and set a September wedding date.

As it turned out, that we even had a wedding was somewhat of a miracle. On September 14, 1968—it was less than a year after we first met, the day before we were to be married, and also Nikki's 24th birthday—I was working with a team at the NIH Clinical Center treating the very first-in-the-world adult ECMO patient. I was so intensely focused on the patient that when I finally looked up at the clock, it was about to strike midnight. I realized with a start that a pre-wedding party had been in full swing at our home, a half mile away. I changed clothes and rushed over, appearing so late that my best man, Robert Sigman, had already volunteered to take my place in the ceremony. I was worried Nikki would be angry, but she greeted me with a broad, beautiful smile, relieved that the main event, our wedding the next day, was not in doubt.

Our party the night before the wedding, Bob Sigman to the left.

We married the next day at Temple Beth Shalom, also a half mile from NIH. We were in the midst of a tumultuous period nationally: the assassinations of Martin Luther King, Jr. and Bobby Kennedy and the 1968 protests that tore apart American cities. Two days after our wedding, her father died. We weathered those storms. We're married still.

September 15, 1968, Bethesda, Maryland.

Three months after we were married, the Navy called my NIH lab, and I was bound for Vietnam. When I told Nikki, I don't remember her objecting. By that point, she had left the CIA and was working at a consulting firm in Washington. While I was hiding from incoming bombs under my cot in a quonset hut in Da Nang, she was watching the terrifying nightly news on TV in our home in Bethesda, Maryland. We reunited two months later in Tokyo, all the more in love. That first overseas trip of mine became a second, and a third—to faraway places in search of adventure, discovery, invention. Nikki would come to recognize the pattern and live with it. My mother had hoped that a wife would keep me at home, and while that didn't quite happen, Nikki kept me feeling I had a home. Marrying her was a stroke of genius on my part.

My mother wanted me to start a family. That wish carried over to Nikki and to me as well, though we decided to wait to have children until I was finished with my residency at MGH. I watched anxiously as David Gabriel Zapol emerged six weeks premature at the Boston Lying-In on May 8, 1972. What a relief it was when he was discharged, healthy if less than robust, a week later. Our second child, Elisabeth (Liza) Bernice Zapol, arrived six years later during my sabbatical in Oxford, U.K., on August 12, 1978. I was there for that delivery, too, although just barely. Nikki's waters broke on a beautiful English summer morning, and we drove to the new John Radclife II hospital high on Heddington Hill. Since it was a Saturday, Nikki's Registrar (doctor) was not on duty. An ample Sister Timson strode into the room.

"I will be in charge today," she announced, and described the upcoming procedures. "Nikki won't need stirrups for her legs. She will be able to use my hips."

I could see the spires of the Oxford Colleges from the window, and Magdalen College would be serving lunch to the Fellows. This delivery could take awhile. I rose from the couch.

"I'll be back after lunch."

"Mr. Zapol," Sister Timson snapped, freezing me in place with her gaze. "I recommend you remain. If you go now, when you return, you will have another child." I sat down.

Sure enough, right around lunchtime, Sister Timson bent over her wide hips, reached between Nikki's legs and brought a beautiful, perfect, crying baby Liza into this world.

Two days later, we drove David and his new baby sister to the pub around the corner from our home in Yarnton. Everyone had a drop of ale.

At home in Cambridge, 2016.

Building a family, building a career—they could seem at odds but in fact they were often complementary, at least for me. Nikki, I am sure, had a different experience. She saw it from the other side. People sometimes ask her what it was like when I took off for Antarctica in 1974. At the time, Liza was years away, and David was two years old. Nikki had just started her first year at Harvard Law School. It was tough. In a law school class of 500 of which 20 percent were women, she was the only woman with a child under five years old. (It wasn't unprecedented—Ruth Bader Ginsburg had managed that feat about two decades earlier—it was still rare.) One of Nikki's classmates, a woman with cerebral palsy in a wheelchair, rolled up to her in the library and said she was amazed at how Nikki could do it all.

Nikki had a system. When the child-care center in Cambridge opened at 8 am, she dropped David off. When it closed at 5 pm, she had to be there to pick him up and wrestle him into his car seat for the ride back to Concord. Her next opportunity to study was after David went to sleep. She compromised her perfectionism and settled for less than Law Review grades. But she made it through. In fact, David graduated from Harvard Law School Child Care Center the same day she graduated from Harvard Law School.

Whether I was in Vietnam or in Antarctica, Russia or Chicago, I was away from home a lot, probably close to a quarter of our married life, until I got sick in 2015. When I was home, I couldn't be relied on to be of much help. In addition to work, I was busy with my hobbies: ham radio, running, biking,

David's high school graduation, 1990. From left: Florence, me, Liza, David, Julia Kaplan (Nikki's mother) and Nikki.

playing violin, reading. I could be pretty intense about whatever I was doing; I've been known to ignore flooding in the basement while engrossed in my violin practice. I rarely changed a diaper. I am not commending or condemning myself—it's just how I was.

Still, we four had lots of fun. We enjoyed cross-country skiing and sledding in the snowy New England winters and swimming in our local White Pond in the summers. We hiked in New Hampshire. David and I sat side by side in my ham radio shack after he got his amateur license as my "harmonic." Both he and Liza, at my urging, attended Hebrew School and, unlike me, at least pretended to enjoy it. Both studied violin, as I had (David and I played the Bach Double Violin Concerto at his Bar Mitzvah). And across the years, I constantly sought ways to include Nikki, David and Liza in my work and lots and lots of travels.

In Boston, Nikki flourished professionally. She had her degrees, one from the Harvard School of Education and one from the Law School, and she applied them to the world we knew, the world of hospitals, serving as a lawyer first for the Dana Farber Cancer Institute, then the MGH. She loved the work. Moreover, she helped me make strides in my work. She was my guide as I navigated the relatively uncharted pathways for bringing a new invention from the laboratory to the marketplace. She taught me how to write scientific papers, warning me: "If you can't write, you can't think." When Harvard sponsored cruises in the Antarctic, Nikki and I lectured as a team. I regaled the passengers with my seal research; she opened their eyes to the important and unique Antarctic Treaty system, a topic she explored while in law school. Together, we enjoyed special relationships with the usually hardy and accomplished tourists who make these adventurous crossings.

Life was satisfying. But I don't want to rationalize it too much. I lived a life specifically designed to push me further, toward new discoveries. Those long stretches where I was absent from home, particularly when the children were young, could not have been easy on anyone. And it wasn't particularly easy when I was home. As I said, when I looked forward into life before I met Nikki, I didn't think of the tradeoffs I might have to make. The same held true during marriage. In a way, that was all I knew. When I was growing up in Brooklyn, my father traveled often. When he wasn't traveling, he was busy working to make money. That was the way it seemed to be with my friends and their parents. I don't remember having personal talks with my father, nor having him ever asking me questions about my life: *What did you do today? What was that like? Who did you meet? What would you like to do tomorrow? Why?* Now I realize that I almost never had those conversations with my own children. Neither parent mentored me in mentoring children. When I was home, I played second fiddle to Nikki, she really was the soloist. I had confidence in her to hold things together.

When I watch the way that parenting has passed through to the next generation, I see that things have changed, a lot. The way I was as a father is nothing compared to what I see from David or Liza's husband, DT. I'm ten percent of them as a father.

There is no knowing now whether I could have done it differently. I realize I didn't try and I realize I didn't feel I had to. Nikki gave me the freedom to work hard at what I loved, and also to have a precious family as my home base. I hope she, David and Liza understand how grateful I am to them, that I truly believe I could not have done what I did without that support. I know they appreciate the joyful times we have spent together, perhaps as much as I appreciate them. And I hope a spirit of adventure and promise and doing good stays with them for the rest of their lives.

CHAPTER 11
Doctor Without Borders

* * *

OVER THE COURSE OF MY life, I have been all around the world, in other countries, in war zones, at the poles. Each of those trips left an impression on me, left me thinking about the ways that people, different everywhere, were also the same, how if you looked hard enough you could always find kindred spirits who were also devoted to the principles of scientific discovery and medical intervention. One of the strongest impressions came back in 1979. It was toward the end of the Carter Administration, which came at the end of a tumultuous and confusing decade for America: it had started with Nixon and Vietnam, passed through Watergate and the energy crisis, and been defined largely by the Cold War, the seemingly permanent opposition of the United States and the Soviet Union, as two global giants facing each other uneasily. Each viewed the other with suspicion, and dialogue and outreach were rare. That's why the phone call surprised me.

It was from Dr. Henry "Hank" Bahnson, then Professor and Chairman of Surgery at the University of Pittsburgh Medical Center (UPMC). I didn't know Hank at the time, but he had combed through the list of grants given by NIH to find a doctor who specialized in treating ARDS who would be willing to travel to Moscow.

Moscow?!? My mind boggled. Who was so important in the USSR that they would request an American doctor? Some Soviet bigwig must be in real trouble.

That was the case, in a sense. Through various international exchange programs and scientific endeavors, Hank had come into contact—and struck up a personal relationship—with a Soviet cardiac surgeon, Dr. Vladimir Ivanovich Bourakovsky. In addition to his position at the UPMC, Hank was in charge of the USA-USSR surgical exchange program. During the tense period in which all bets were off as to whether a real war would break out between our two nations, scientific programs such as the surgical exchange program were among few lines of diplomatic communication that weren't forms of saber-rattling. Bourakovsky was Hank's opposite on the Soviet side. Friendships moved across all borders.

As it turned out, the unfortunate ARDS patient was not Bourakovsky, but his adult daughter. She presented as a terminal case, unless she was cared for with the proper techniques, medicines and equipment. I had spent the early part of my career working on these things exactly. Hank and I went through it on the phone.

"Zapol," he said, "I've read your papers and NIH says you are an ARDS expert. I don't want an academic, I want someone who can save a life."

"Dr. Bahnson, I can assure you, I spend my clinical days in the ICU, treating patients with ARDS." I didn't hesitate to volunteer.

"How often are you in the clinic?"

"Four days a week," I said.

"Oh good," He seemed relieved, just for a second, but then started in again with the interrogation. "When was the last time you had an ARDS patient with an infection?"

"I'm on the floor now, most of the patients here, now, have sepsis."

"What kind of ventilators are you using"

"Volume-controlled ventilators."

"Can you do it alone?"

"Nothing is done alone, sir, but I'm willing to fly in to help the team in Moscow to see if we can rescue her. I'll mobilize everything we've got here at MGH to support her recovery. If it can be done, I'll do it."

Marina Bourakovsky was a physician-scientist like her father. She had suffered complications owing to an abortion gone wrong, and there were rumors about who the father had been. He was nowhere to be seen when it came time to have the abortion, never mind what care was necessary from the complications afterward. Those complications were both common to many women, and extraordinary given her position in Russian society. Marina developed acute abdominal pain and a high fever, making it impossible for her to hide the fact that something was wrong. Though her relationship with her father was strained, she turned to him for help. She told him about her abortion, which angered him. But he helped her. If he hadn't, and if he hadn't been so very well connected, she would have died for sure.

At first, he followed normal procedure, asking one of his most trusted colleagues, who was also his family doctor, to treat Marina. She was admitted to Moscow City Hospital where her father had many friends. The doctors performed a laparoscopy under general anesthesia, but it didn't go well. During the procedure Marina suffered a cardiac arrest, requiring immediate cardiopulmonary resuscitation, intubation and mechanical ventilation. After she had been revived, they proceeded with a laparotomy and removed an infected fallopian tube containing a fetus. This had been an ectopic pregnancy— the fetus had developed in the wrong place. The laparotomy didn't take long, Marina remained under anesthesia for the entire procedure. Thankfully, there were no additional complications during that operation, but the cardiac arrest was only the beginning of her troubles.

Vladimir Ivanovich visited Marina in the City Hospital that evening, and assessed her condition. He became concerned. As he was a director at another hospital, the Bakulev Institute, he knew he could access better resources for treating his daughter. At that point, Vladimir Ivanovich himself took charge of her care, ordering equipment sent from Bakulev to the hospital where she was being treated, probably concerned about the danger of a patient transfer. This part of the story concerned me. While I certainly understood the impulse to take charge of the treatment of your loved ones, it is rarely advisable. How can you make a clear-headed risk assessment concerning the case of someone you love, especially your child? It's almost impossible. From what I could tell, the shift in control to her father resulted in significant delays in calling me over, delays which appeared to be endangering Marina's life. She stayed at Moscow City Hospital, where she remained unconscious and began to develop progressive acute respiratory insufficiency on the second or third day after her surgery. I remember looking at the chest radiographs that had been taken in those early days, which revealed many non-ventilated lung regions, which could have easily spelled death in an older or less lucky patient. Eventually a tracheostomy was performed to make her care easier. Antibiotics were continuously administered intravenously, without any effect.

During her treatment, each day more and more physicians became involved. Vladimir Ivanovich listened to each of them, considered their advice, but didn't actually alter treatment. In part that was because the advice was profoundly unsound. I heard about it later from Dr. George Falkovsky, a fluent English speaker and eminent pediatric cardiac surgeon at Bakulev. George had just finished interpreting a call to Hank Bahnson during which Vladimir Ivanovich Bourakovsky raised the possibility of bringing in an American specialist to treat Marina. Vladimir Ivanovich picked up the phone and called the hospital to check on Marina's status, and George overheard the following exchange:

"All right, good, try it… wine in the rectum? And how would you administer the enema? Through a tube? Good. Only not too dry, better use the sweet Georgian wine, the soft one, perhaps Kagor? Yes…"

He hung up.

George looked closely and intensely at him. "Vladimir Ivanovich," he said, "Have you ever seen a doctor in the USA who would inject wine into a patient's ass?"

This was the thunder-strike needed for Vladimir Ivanovich to snap out of his stupor. He immediately called back the doctor at Moscow City Hospital and shouted into the phone: "Have you gone completely crazy over there? What wine enema? What are you doing to my daughter!?" It was also the moment that sealed the deal for my visit.

Even with the decision made, the intervention of many prominent Russians was required. One of them was Evgeni Maksimovich Primakov, who would later become Prime Minister of Russia. At the time, he was a Chief of the Institute of World Economy and a very important member of the Central Committee of the Communist Party.

After Primakov pulled some strings, and the Georgian wine stayed in the bottle, I got the call. I was anesthetizing a cardiac patient in the White Building Operating Room 8 at MGH when I was paged to speak with a man who identified himself as Dr. George Falkovsky, translating for Dr. Bourakovsky in Moscow. I had some sense that the call would be coming. I had spoken to Hank, of course. Still, it wasn't every day that I received an urgent medical request from Moscow.

George rolled his Rs a bit. His voice was full and rich and I felt like I could believe what he was saying. Perhaps it was because his English was so good. Perhaps it was an air of trustworthiness created by the smoking and Russian vodka, things he would introduce me to later.

"Warren," he said. Maybe there was a third r in there. "Warren, she is a young patient, Marina, she is dying of shock lung, after gynecologic surgery during which she had experienced a cardiac arrest." Shock lung was a term that they used for ARDS. Marina had severe hypoxia, renal failure and was comatose. "Warren, what should we do, what would you do if you were here?"

"George, thank you for the background. I can only speculate from here, but based on what you told me, I suggest administering PEEP: positive end expiratory pressure. That may help. And of course, we'll need to find the cause of her deterioration. My instinct is that it could be pneumonia or possibly abdominal sepsis."

We discussed for another minute or two, then the call ended quickly.

"Thank you, Warren, thank you," George said, and hung up.

After that call, I remember talking it over with the surgeon in the operating room. As much as I was game for an adventure, I thought that George might just be able to handle the situation. I tried to imitate his accent. "Warrrrren". With that voice like velvet, of course he could command the situation. But the next day George called again.

"Warren, she is worse today. The situation is going downhill. I've spoken with Vladimir Ivanovich. Please come quickly to Moscow. We need your help." he said. When I thought about this poor woman, her case reminded me of the women I had treated who were suffering from similar operations gone wrong during my internship in Boston. I couldn't give a hoot about the Cold War, the strings being pulled from who knows where. I just knew that if there was a chance that my knowledge could save Marina when no Soviet doctor could, then I would take it.

I called Nikki in Concord. She had long learned to expect the unexpected from me. At least I wasn't going into an active war zone again. She did press me with a few pointed questions like: "Who knows about this?" Protective questions, triggered by her own experience working at the CIA, and her experience as a lawyer. She knew I'd made up my mind to answer this call, she just wanted to make sure it was safe. She offered to pack me a suitcase with some warm clothes for winter in Moscow and be ready to bring it into the hospital in case I had to leave very soon.

I began medical preparations immediately. I asked Dr. Henning Pontoppidan to help arrange for the MGH to donate a volume ventilator, and I asked Dr. Mort Schwartz of MGH Infectious Disease to advise me about an antibiotic. Mort advised me to take BBK8, an experimental medicine that had recently been approved and today is called amikacin. This new antibiotic was good for gram negative bacteria and in all likelihood not available yet in the Soviet Union, so the patient was very unlikely to have an infection that had developed resistance to the drug. Preparing for the worst-case scenario, I took an ECMO spiral coil oxygenator and cannulas. All this was packed in bags for the journey.

That night I received a call from the U.S. State Department telling me not to go to Russia, warning me about traveling without a visa. Would I be imprisoned at Sheremetyevo airport upon arrival? I talked it over with Nikki late into the night. I had been assured by George that I would be taken care of before I got on the plane. We decided that I would not board the plane for Moscow the next day without a visa.

I arrived in Kennedy Airport late that afternoon as planned—certainly the only traveler with BBK8 and ECMO in my bags, and also the only one to be met by the Russian consular official George had promised. His first comment was one of disbelief: "You look too young to be Dr. Zap-ol," he said. I assured him he had the right man. He assumed a grave expression that told me he had been briefed on the nature of the situation and its urgency. He stamped my passport with a visa, handed me an Aeroflot ticket to Moscow and told me my bags would be loaded onto the plane. I seated myself in tourist class but soon was ushered up to first class.

Despite the extra comfort, I barely slept during the long flight to Moscow. To tell the truth, I was a little panicked. I was by no means a novice traveler, but this was completely different from pleasure travel, or conference travel. There I was waltzing off alone to my nation's greatest geopolitical rival. But then I closed my eyes and thought of George Falkovsky's warm and reassuring voice, like the voice of a friend, rolling his Rs. That made the trip feel safer.

As we landed, my anxiety began to build again. What would happen when I landed? It had been a cold winter in Boston, and the image sprung to mind of getting stranded in some Siberian gulag for who knows how long. I overcame my fear by reminding myself that someone's life depended on my composure and my expertise.

On the runway in Moscow a stairway rolled up to the plane, the door opened, and along with a blast of cold air a smiling man came into the plane. "Who is Warren Zapol?"

"It's me!" I said. "You must be George Falkovsky!"

I'll never forget how both of us grinned from ear to ear. He was just as warm in person as on the phone. I knew instinctively that he was a man I could trust, and I was right. I count him as a lifelong friend to this day.

George grabbed my bag from the overhead and beckoned me to follow him down the stairs from the plane. There he introduced me to Bourakovsky, seated in a beautiful, luxurious Volga limousine. Vladimir Ivanovich Bourakovsky lived up to his romantic sounding name. He was a large, imposing man with white hair and a cherubic kind face. He also greeted me warmly. The chauffeur drove us from the plane to the main airport terminal, then escorted us to the VIP hall, where he took my passport and disappeared to have it stamped with my entry visa.

Suddenly, a short man with a mustache appeared at my elbow. He had a dapper hat pulled down to just over his eyes. He introduced himself as a *New York Times* reporter. "May I interview you?" He asked. "Please... Do you know why you are here? It is an extraordinary situation, an American doctor, here!"

"I've been told that there is a sick patient, a young woman, who needs immediate help. That is all I know," I replied.

"And who is this woman? Is she a relative of a big party member?"

"I don't care about that. She is sick."

"What do you think, is there a chance she could make it?"

"As far as I know, her condition is pretty serious. I can't tell you anything before I see her."

He repeated his question: "Any chance she will survive?"

"If she survives, I'll be glad to talk to the press. Until then, I have nothing to add." I cut off our conversation. How this reporter found out about my arrival and where he came from is still a puzzle to me. I had thought my mission was secret. I assume that the American embassy knew, so perhaps someone there tipped him off. At any rate, I had not given him enough to run a story right away, though one would eventually come out.

The reporter vanished. Our driver reemerged with my stamped passport. Just like that I was admitted for entry into the Soviet Union. Though it was only afternoon the sky was dingy, and George talked the entire slushy way to Red Square. He told me that he had arranged everything for my stay, right down to a luxurious room at the Hotel Metropole, with a spectacular view over the Kremlin and Red Square. The view was indeed impressive, as was the hotel itself, but I took little notice. I wanted to get to the City Hospital straight away and meet the patient.

When we arrived at the hospital, I immediately set about organizing the team caring for Marina into a mini one-bed ICU group. I relied upon George not only for English interpretation but for facilitating just about every task before us. Dr. Lado Meskishvili was our clinical mainstay, and I asked for a large piece of paper and fashioned an MGH-style ICU Care Sheet on which staff could enter Marina's vital signs, drugs, and arterial blood gas measurements for easy viewing.

I'll never forget the look on the staffs' faces when I asked them if their Dräger ventilator had PEEP. As I had already conveyed in my call with George while I was in the operating room at MGH, PEEP is what we would have used back in the States—it is important for a number of reasons, but mostly because it keeps injured lungs supported and open at the end of the breath. When you breathe out all the way, you are usually just fine. But if your lung collapses, the "positive *pressure*" of PEEP can help to keep the lung open at "the end of expiration." But the blank stares that greeted my question set off a lightbulb: the technology had not yet been introduced to the Soviet Union. I would have to make my own. "Where can we get a 10 liter can of water?" I asked.

George looked quizzical, then smiled, turned to the staff, and conferred with his colleagues, and someone was sent off to procure one. We filled the can with sterile saline, disconnected a line of the ventilator, put it into the can, and grabbed a marker and drew a line 10 cm above the hose inlet. The expiratory gas would pass out through the fluid, creating a calibrated expiratory pressure. On that day in early March, 1979, we constructed the first ventilator with PEEP in the Soviet Union! It was not just innovative, but important—if Marina's lab values improved, we could just reduce the level of inspired oxygen and keep the end expiratory pressure in her lungs constant while tracking her lab readings on the ICU Care Sheet to make sure she was responding well. The acute respiratory care was set up now, and though we didn't see immediate improvement, it was critical to her life support that we be able to adjust these parameters before we were forced to use ECMO.

With the ventilator now working, I requested a search for a pressurized air source to dilute the 100% oxygen Marina was breathing. It can be damaging to patients, and we rarely give 100% oxygen for extended periods today. Before I left, I also stopped all the drugs she was on. I asked for her blood and sputum to be cultured to see what was behind the infection.

Finally, I was done with the first phase of care. I was ushered out of the hospital room, back down to the front of the hospital, where the Volga limousine was waiting to bring me back to my hotel. Through the window of my room, I looked up into the darkness of the falling snow, into the lights of Red Square and the Kremlin, and before long I had fallen into my first restful sleep in days.

The next morning, I met George at 8 am. The prestige vehicle they had picked me up in at the airport was nowhere to be seen. Instead, George drove me in his much more humble car, a Zhiguli. Even that said something about George's stature in society—personal vehicles were rare, and every time George parked the car he removed his windshield wipers and carried them with him to prevent them from being stolen.

On Day Two, we set about trying to find the cause of Marina's illness in earnest. I believed the destructive process in her lungs was related to sepsis and pneumonia, and that it was critical that we kill the responsible bug. The antibiotics the Soviets had given Marina weren't working, which was why I had stopped them and started her on BBK8. Our new strategy was to perform a blood culture every three hours until we could "catch the bug" and be sure that we were on the right course with our new and different U.S. antibiotic. The lung ventilation regime using 10 cm water PEEP levels was constant, and a portable chest X-ray was taken each day to detect any pneumothorax (when the lung ruptures and air leaks into the pleural space), which would be catastrophic at high PEEP levels.

It was then that the differences in standards at Soviet and American hospitals became even clearer to me. They could easily collect venous blood for a blood culture, but it took forever to grow bacteria in Moscow and learn the result. There were many old antibiotics, and surely bacterial resistance had evolved to all of them. The Russian nutrition that was given to Marina through her IV was a strange, milky substance that I didn't trust at all. I asked to be connected to MGH in Boston and after an hour on the phone with Mort Schwartz, he said he would support regular deliveries of BBK8 twice a week via Aeroflot (tucked in the cockpit) to Moscow. My wonderful administrative assistant at the time in Boston, Shirley Barry, organized the logistics. I asked her also to provide an IV food, or hyperalimentation solution from Boston Shriners Hospital. As we couldn't reliably store the solution in Moscow, we planned for Aeroflot deliveries from Boston every second day. Shirley was even able to provide us with much-coveted Marlboro cigarettes and MGH surgical uniforms. And all of this, across an ocean, a continent, and through the Iron Curtain that separated the U.S. and USSR.

We kept hunting the cause of Marina's sepsis as though it were a fugitive. George and I went to the microbiology lab, which was located in the basement, to find a calm, soft-spoken bacteriologist, who seemed completely dumbfounded by the American and his translator in the lab. I asked questions through George. "Did anyone give you Marina's blood in order to perform blood culture?"

"Yes, once."

"And when was that?"

"I think it was five days ago."

"And afterwards?"

"Nothing afterwards."

"So, is something growing thus far?"

"Well, nothing so far. But, you know, the culture medium we have is kind of slow. Of course, if you want to take a look… My sense is that one plate looks suspicious."

He was right. On one plate something was growing. "Did you check to learn if it's sensitive to the antibiotics we gave her?"

"Well, it's still too soon, you can barely see anything, but we will in a couple of days. By the way, I told the doctors that this blood culture could be positive, but they didn't listen. Anyway, I can't tell for sure right now."

"Could we please take a look at a tracheal smear? You performed it, right?"

"No. But if you could do it yourself, you are welcome to use the microscope, stains and all of the equipment we have."

"Thanks. We'll be right back, just give us a couple of slides." We weren't able to see anything when we smeared out her blood or sputum.

It was clear we needed modern microbiological capabilities. It seemed crazy, but I had an idea where to look. I knew that Russia had biological warfare scientists, we all did, it was an open secret. My idea was to invite them, the warfare scientists, into the hospital. My request was met with raised eyebrows, to be sure. But somehow, maybe Primakov was involved, later that day members of the Moscow Chemical Biological Warfare Group appeared in white gowns and what looked like tall white chef's hats. My jaw dropped. Their day job may have involved cooking up poisons, but here they acted as benevolent scientists, taking Marina's samples. A few days later, they reported that indeed *Pseudomonas* had been cultured from her blood and sputum. We also learned that Marina's particular bugs were sensitive to BBK8 in the lab. This may have been my most important medical advance in her case.

I visited Bourakovsky in his office on that day. He was with his pathologist, Leonard Crymsky, whose arm Burakovsky held up for me to see. It had a number on it, a concentration camp number. Bourakovsky immediately assured me that I had nothing to fear. He had hired Jews, he said, and he added that I would never be followed by the KGB. It was a disconcerting moment that somehow reassured me.

I told Bourakovsky and George that I needed to find the anesthesiologist who had been in charge of Marina during her surgery. Something had clearly gone so terribly wrong then. They located the man and I asked him to sit with me and review every chest X-ray of Marina from the very first day of her illness. On the first chest radiograph I quickly noticed a long shadow, descending from the middle to the lower lobe of her right lung. Was this an aspiration pneumonia? The anesthesiologist said he didn't think so, perhaps he hadn't noticed, or perhaps he didn't want to draw attention to a situation where he could be found to be at fault with such an important patient.

"That guy is lying," I said to George when he left, "I can see it on the X-Ray."

I left the hospital the second day in a gloomy mood. The many attempts to save Marina had proved useless. She remained unconscious and appeared to be dying. We were all tired and exhausted, and jet lag was taking its toll on me. Still, I was unwilling to accept the prospect of failure. I was determined that Marina would live.

That evening, I accompanied George to his home, met his wife Elka, and their daughter Olga, and enjoyed a meal, complete with pickled mushrooms that they had gathered near their dacha outside Moscow. It was a rare moment of relaxation during those first few days. Is it strange that I remember the pickled mushrooms? I have foraged for mushrooms on Cape Cod with George, but haven't had them pickled since that night in Moscow.

The next morning I entered Marina's room to encounter a changed mood. BBK8 seemed to be working. Marina had no chills anymore. Her body temperature was still high, but her skin color was healthier, the blue even edging towards pink, and her kidney function was improved. On top of that, the shadows on her chest X-ray began to disappear. This marked improvement worked miracles on group morale. Doctors and nurses began to smile more and more. Vladimir Ivanovich was still a regular visitor, and I could tell that the smiles on the faces of the staff were as much a reflection of their relief for Marina's father as for Marina herself. Somehow, this Cold War-era medical supply chain, from MGH in Boston to New York onto an Aeroflot plane to Sheremetyevo to Moscow City Hospital was working well enough.

From then on, my stay in Moscow became much more pleasant. Most of the doctors saw me as a friend now. They knew there was no hidden or malicious intent in my actions, despite the general paranoia that pervaded every level of Soviet society.

That night Dr. Bourakovsky and Yevgeny Primakov took me to a Georgian dinner restaurant called Aragvi, where we toasted Marina's recovery with vodka and Kindzmarauli, Burakovsy's favorite Georgian wine. Between sips, I was informed that it had been Stalin's favorite. (I had to wonder whether this was also the wine that had been recommended for Marina's treatment. It didn't seem appropriate to ask.) We gorged ourselves on caviar and then devoured Georgian pizza—a cheese filled bread called khachapuri. I was wary, though, that our ebullient optimism was premature. Though she had come back from the brink, she still had renal failure and fixed dilated pupils. What if Marina survived, but only in a vegetative state? At one point in the night, Burakovsky turned to me, the smell of fruit liquor on his breath, and asked if I thought Marina would survive. I told him the truth, that it was still touch and go.

Over the next several days Marina continued to improve. Her arterial blood oxygenation level rose and her urine output increased. The morning of the fourth day, all IV sedatives were stopped so Marina could wake up. There were three of us in Marina's room: Vladimir Ivanovich, George and myself. Vladimir Ivanovich was shouting at our patient.

"Marina! Marina! Can you hear me?"

To my amazement, Marina opened her eyes abruptly and nodded. She looked quite conscious. "Are you in pain?" Marina shook her head no. "Do you realize where you are?" She nodded affirmatively. She looked scared but alert. At that point, her excellent thoracic surgeon, Michael Israilovich Perelman walked into the room. "If you spoke that loud to the ears of the bronze statue of Lenin in Red Square," he said, "It would open its eyes." The tears welled in our eyes. Marina was a living, breathing woman, a human being who wanted to live.

It was miraculous. And yet Marina's life was hanging by that thread of hope keeping her from the abyss. She was so weak, there was no room for further complications. I remember once telling George my rather simplistic but ultimately effective approach to intensive care—"The basic principle of intensive therapy is: When you are winning, don't change anything."

We didn't. Marina's condition and, most importantly, her lungs continued to improve. She was able to answer all the questions we asked her by either nodding or moving her eyes, and she was becoming more active almost by the hour. Though she still couldn't speak because of the tracheostomy and mechanical ventilation, she was very attentive, observant and interested in everything that was going on around her. It was a true joy to behold.

Toward the end of the week, George made the wise suggestion that we take a little R&R for ourselves. He invited me again to his home, to indulge again in their wonderful Russian hospitality. This time I got to taste Elke's home-made strawberry jam on toast with tea. We spent an evening listening to Chuck Mangione records, no doubt smuggled into the USSR at great difficulty and risk. I found myself forming a close bond with George and was honored when he opened up and told me his story.

George could speak English because his father was an American, making George a true rarity in communist Moscow. His father had been a committed communist in New York and a writer for the *Daily Worker*, a newspaper generally reflecting the views of the U.S. Communist Party. His father had left America, traveling first to Germany (where he worked as a miner) and then to Russia, where he met George's mother, a descendant of Scottish socialists. George's father started the *Moscow News*, an English language news service. With wide eyes, George told me how his father, disillusioned and threatened by Stalin's purges, escaped Russia in 1939 by skiing alone cross-country from Leningrad to Helsinki, Finland, leaving George with his mom. Then his dad joined the U.S. Army and fought the Germans in Italy, before returning to New York and marrying again.

Before I left, I had the chance to meet George's mother Rose, who was living in a Moscow apartment. She repeated the story, without a hint of bitterness. And months later, after returning to the USA, I met George's father, who was living in a retirement home in Manhattan. Along the way I learned that George had an American half-brother, Paul Falkowsky—a distinguished oceanographer who would eventually become my scientific collaborator as well.

* * *

Marina still faced two major hurdles for long-term survival. The first was providing her with appropriate mechanical ventilation. Back in the 1970s, even the newest and most expensive ventilators in Russia were imported from Germany, and even *those* were not able to provide assisted ventilation for the process of taking her off the ventilator. As Marina improved, she was transported in a special ambulance, escorted by black Volgas with flashing signal lights, to the ICU of the Bakulev Institute, while an Aeroflot freight carrier-plane with a modern Emerson mechanical ventilator from MGH was on its way to Moscow. By that point, she was almost able to sit upright in her bed and to swallow fluids, while still being supported on the ventilator. It was far easier to take care of Marina in the Bakulev Institute than at the City Hospital. When the Emerson ventilator arrived, Marina was switched to spontaneous ventilation.

The second major problem was her ongoing sepsis. Despite the antibiotic treatment, she developed fevers to 39°C several times a day. We were putting out fires in a war of attrition against an infection we couldn't seem to pin down. We frequently obtained blood cultures, but the lab still didn't provide a definitive answer. I was constantly torturing George with my questioning: "Where in her body is the bug coming from?"

Eventually, we decided that an exploratory laparotomy might be helpful. My theory was that someone had overlooked an abscess in her abdominal cavity during her first laparotomy and it was maintaining her infection.

We called a meeting of twenty physicians, including chiefs of surgical departments and their representatives from the First City Hospital, Botkin's Hospital and the First Medical Institute. Not one could answer the question of why Marina was still septic. The mood in the room was tense. Despite the fact that our success had won over the doctors on the team around me, to others I was still regarded as a meddling stranger. George introduced me to a gentleman he said was "probably the greatest surgeon in the Soviet Union." The surgeon spoke derisively and dismissively to me. "What does the American say? No—I think it's wrong! There is nothing in her abdominal cavity."

I insisted, and another laparotomy was performed by the chief of surgery, a fine general surgeon. In fact, the surgeon didn't find anything suspicious in the abdominal cavity. George and I were in the operating room during the procedure. It was only then, seeing it for myself, that I was convinced that the source of sepsis was not in Marina's abdomen. I had been wrong and it felt like something of a setback. I defended my position to myself and to George as being thorough, and I defend it as being thorough to this day. But I knew it had cast doubt on my medical judgment in front of others who were already doubtful of me.

At that point, another problem came up, and again no one was prepared for it: Marina just didn't want to see any visitors. During our attempts to have a conversation with her, she would just close her eyes and fall asleep. The poor woman was exhausted, and she may have had some level of depression as well. She needed something doctors could not give her. I had gathered enough about the Burakovsky family by then to know what to do. I asked for Marina's sister Lena to come and visit her. Lena was living in Jordan, where her husband was working as a diplomat in the Russian embassy. Lena flew to Moscow, where she played an important role in Marina's recovery. Lena was a very warm and happy person. She smiled a lot, and was always in a good mood. She was kind and effervescent. Marina immediately improved. She began to eat, to smile, and even communicate. The only reason Lena would ever leave Marina was to get some sleep. I remember her sitting beside Marina and holding her hand continuously. She needed love, that was the most important medicine for her then.

Three months after visiting Marina in the hospital, Lena died in a tragic car accident in Jordan. At Lena's funeral Marina declared, "She is lying there in her coffin instead of me." Marina was convinced the fates had been aiming at her.

With her road to recovery more secure, I was able to ease off a little. I spent the first part of each day at the Bakulev Institute and afterwards I was taken to visit museums, concerts and restaurants. Gradually, things returned to normal. As I remember, it took us several days to convince her that she would be fine breathing without mechanical ventilator support, and that it was the only way to get rid of the tracheostomy tube and speak to us. I was working hard, now as her psychologist. Lena helped a great deal, holding Marina's hand and explaining to her the necessity for spontaneous breathing. Finally, Marina agreed, and was slowly but successfully disconnected from the ventilator. At first, we did this during the daytime for brief periods and later while she slept. We decided, finally, to have a brief press conference. Marina wasn't receiving any antibiotics now, was eating regular food, and there was no need for continuing IV food; she was off the ventilator and breathing spontaneously without a tracheal tube. We felt sure of her recovery.

The victory press conference was given in the conference hall of the Bakulev Institute with many journalists asking deep questions. I tried to tell the whole story, why and what I had done, emphasizing the help I had received from

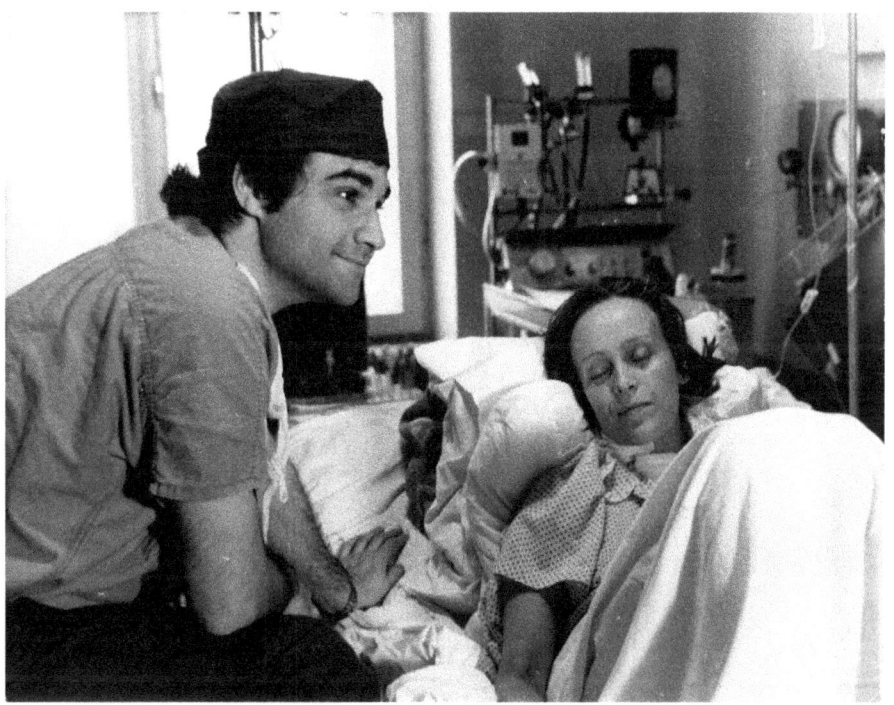

At Marina Bourakovsky's bedside.

everyone during Marina's treatment. I still have the photos taken of this conference where George, Vladimir Ivanovich, and I were sitting in front of the
journalists. George and I were talking, looking in different directions, while
Vladimir Ivanovich was sitting near us, looking concerned. He was leading
the conference, picking the journalists from around the world who would ask
questions, including the reporter from the *New York Times* who had found
me on my first day at the airport. Years later, looking at this picture of himself
with George and me, Vladimir Ivanovich would smile and say "Two rabbis…"

On March 16, I celebrated my 37th birthday. George and I went to
Vladimir Bourakovsky's flat in the famous "House on the Embankment" near
the Kremlin, a beautiful but eerie site of Stalinist purges. Vladimir's gracious
wife, Lilly, provided a fabulous spread. Among the special guests was again
Yevgeniy Primakov, who at that very moment was planning the Soviet inva-
sion of Afghanistan. Georgi Arbatov, head of the Institute of the USA and

Press conference in Moscow with George Falkowski and
Marina's father, Vladimir Bourakovsky.

Canada, was also there, another powerful guest. After a group toast to Marina's recovery, my birthday, and the good health of my family, Primakov turned and looked me in the eye. "Zapol," he said, "I thank you for coming to Moscow and saving the life of a Russian girl. When the generals want to push the button, I will remind them that you came to Moscow to save the life of a Russian girl." Others laughed, but his tone was entirely serious. What he said made my hair stand on end, as it does every time I relate this story.

On my last day at the Bakulev Institute, I withdrew a blood sample from Marina's radial artery and performed a blood gas analysis. The arterial oxygen saturation of her hemoglobin was 96%. There was nothing to worry about, and both she and I could now safely go home.

But this was not to happen before some very Russian-style celebrations. As a parting thank you, I was escorted to a fabulous performance of The Magic Flute at the Moscow Opera and Bolshoi Ballet. Prof. Bourakovsky gave me a sable fur hat, as well as a beautiful old pocket watch. I remember asking: "Was this made before the Revolution?"

"Of course," was his response. "Everything good was made before the Revolution."

Returning home after all of this was a bewildering experience. Being in Moscow for several weeks at the height of the Cold War was something that most Americans could hardly imagine: the differences ran deeper than windshield-wiper thefts or the way the food tasted. It really was like stepping into another world. Getting pulled so deeply into that world was something I could never have imagined. I returned home to 1970s American life with these memories. I was exhausted, but proud and relieved. I had just come back from another world, where we had pulled a woman back from the threshold of death.

I was more than a little dismayed, when, upon my arrival back at MGH, I was presented with bills for the drugs and supplies, including the Emerson ventilator that I had shipped to do the job for me. Of course, I knew that was probably coming—and perhaps there might have been a way to make the

Soviet government pay—but I was too worn out to go the diplomatic route. Fortunately, the bills were paid by the drug company that sold BBK8, Marina's life-saving antibiotic. They had received a great deal of positive publicity from this adventure.

Throughout these years George and I have remained very close friends. It was hard to imagine how we couldn't after all we had been through. I visited him and his family in Moscow several times where I was always treated in the best and warmest way. He moved to Israel not long after my visit, and then to the U.S. with his wife Elke. Later his daughter Olga immigrated and became a radiologist. George worked briefly in my lab at MGH, then moved to Long Island where we recently celebrated his 80th birthday.

Vladimir Ivanovich was likewise a devoted friend and host to me. A year after my medical mission, he and his wife Lilly and many others who were part of this extraordinary adventure treated me and my family to a fabulous vacation in Moscow, Tbilisi and Kiev. I have such warm memories of parties and animated gatherings everywhere we went. It's memories of friends like those, untethered from the problems of politics that really matter: they're the ones that stay with you.

And Marina completely recovered and went back to work a couple of months later. Now she is a grandmother and is helping to raise a wonderful and smart granddaughter.

✳ ✳ ✳

My return trips to Russia were trips of friendship, not of work. But it wasn't my only international life-saving mission. A few years later, in 1985, I received another call, this one from a Brigham-trained surgical colleague, Dr. Jose Eduardo Cunha, then living in São Paulo.

He told me that the President-elect of Brazil, 75-year-old Tancredo Neves, was dying of severe acute respiratory failure. Could I help the president? I agreed immediately—how could I not?—and again secured some medical equipment, some experimental drugs and schlepped an ECMO machine to JFK airport in New York, while trying to dodge the press. Still, the *Financial Times* caught me. "Dr. Zapol," the reporter said. "We hear you're going to Brazil."

"Yes," I said.

"We also hear that the military has shot Tancredo Neves, and we think you're going to try and take care of him."

"That's not what I'm hearing," I said. They remained suspicious and pressed this theory. I remained steadfast. "I have no evidence that he's been shot in the abdomen," I said. "It doesn't sound like they're trying to kill him. But I'll tell you what I find."

I landed in Rio. I had received explicit instructions from Jose Eduardo to avoid the Brazilian FBI, not even to answer to my name when called. I didn't identify myself and went from gate to gate, switching to fly to São Paulo. When I landed in São Paulo, I was not so lucky. The Brazilian FBI spotted me immediately. They put me in a car with my equipment and we drove through darkened streets to the Instituto do Coração hospital. Groups of people were gathered on candle-lit street corners, praying for the life of Tancredo. As I learned, he had been hospitalized in Brasilia on March 14, 1985 with a severe abdominal infection. He had a benign small bowel diverticulum that had twisted and its blood supply was cut. Unfortunately, he pulled his nasogastric tube out to speak on national TV, which opened his abdominal wound and started a bleed from the stump of the diverticulum.

At his next surgery, the entire parliament of Brazil had insisted on being present to watch. They didn't wear masks, or gloves, or anything. Conditions were filthy. On top of that, he needed blood transfusions but there was an apparent shortage. That led to low blood pressure and the low flow caused kidney failure. Three or four operations were performed in Brasilia before Neves was transferred to Instituto da Coração in São Paulo.

For a week I worked with Jose Eduardo as my translator, creating a bond like the one with George. Our families are close to this day. I saw Neves in an ICU bed with no one around. He was given a whole floor. But all that attention didn't mean that the team was helping him. When I arrived, the ventilator was set so high that it was literally blowing up his lungs. On top of this he was septic, on antibiotics, and was doing dialysis for severe renal failure. The U.S. Embassy offered for me to transport him to the U.S., but I realized after a few days that such a move would not help him. I did not even use the ECMO machine I had brought. Instead I cooled Neves to 86°F, which is what we sometimes have to do for patients at this point, and added a drug, L-3,4 dehydroproline to block fibrosis and finally I reduced inspired oxygen levels. We tried hemodialysis. None of it worked.

There were perks to treating a president. Many evenings we had wonderful Brazilian jazz singing and guitar performed by young surgeons in the hospital. Music was in the air despite the somber mood. The President's sister was a nun, and so they cleared out the next floor as well, where continuous prayers were said for the President's survival.

After a week, we were not making any progress. At that point, a friend of mine by the name of Bill Knaus had come up with something called the APACHE score [Acute Physiology, Age, and Chronic Health Evaluation], a disease-scoring system. Using that system, we could assess the probability that someone would survive an acute illness. We filled in the data and totaled it up. There was a 99.99 percent chance that Neves was dead. We reviewed his progress in the morning, discussed what we might do, and that afternoon I said the unsayable: "Perhaps it is time to give up on medical care."

It was a difficult decision to make, and I didn't realize it wasn't a decision that was for me, or any other doctor to make. The military attaché assigned to me spoke up. "Neves can not die today," he said. "It is Tiradentes Day. This is a Brazilian national holiday in honor of a martyr, a leader of a failed independence movement against the Portuguese in 1789." The man had been a dentist, which gave the day its name: Tira dentes, or tooth puller. "If Neves were to die today, it would not be good."

Jose Eduardo explained to me that the military was concerned that Neves' passing could be taken as a call for revolution against the military. I could hear the people in the street singing. We walked out onto the balcony and looked down into the street. People were cheering for us. I knew he wasn't going to make it another 24 hours, but these people certainly did not.

"Okay," I said. "I can stay with the President until the day is over, keeping him on life support. Then when the day has passed, he can pass as well."

I returned to the room and sat with Neves, watching the air flowing in and out of his lungs, listening to the song rising off the street. Toward the end of the day, my Naval Attache returned. "It is now officially acceptable to allow Neves to die naturally, whenever it might occur," he said.

And so it passed that Neves died on Tiradentes Day, with me as a witness. The Attache asked me to do an independent autopsy because I was a medical expert from outside the country. I told them what I found: no bullet wounds, no bullets, no shattered bones, no evidence of foul play. Natural causes.

The crowd in the street swelled when the word of Neves's death spread throughout the city. A mob collected outside the hospital. There were maybe half a million people, maybe more. They were chanting his name, "Tan-cre-do! Tan-cre-do!" and there were fires in the streets. The Brazilian FBI snuck me out of the hospital through the basement.

For the next few days the country was in mourning, and then national affairs resumed. The Vice President-elect, Jose Sarney, became the new president of Brazil, and though he was greeted with some initial optimism, in time the crippling debt crisis, chronic inflation and corruption hobbled the country.

For me, it was deeply humbling to be charged with the care, and to oversee the death, of Neves. He was a historic figure, but no matter how important we each may have been in life, when the time comes, the time comes.

CHAPTER 12

Uncle Sam

* * *

As a traveler and adventurer my whole life, I've made many close friends that I consider family. They come from just about everywhere in the world, which means that I have felt at home all around the globe: in Copenhagen with Jesper, in Bavaria with Konrad, in Antarctica with Bob. Similarly, Nikki and I try to make my international group of fellows feel welcome in our Boston home, as part of our extended family. I call them my intellectual offspring, much to my biological childrens' consternation. These offspring have returned to China, Japan, Italy, Germany, Argentina, Russia, even across the river to Charlestown. I imagine a map with pins in every continent.

Yet somehow my own muddled beginnings on Atkins Avenue pinned me to a place and time, and I have never been able to come unstuck from it. To be honest, I don't really want to. I'll always be my mother's son. Or rather, my mothers' son: both of my mothers. I can't forget either one or detach from either one's memory. But after the secret of my birth was revealed, it was almost too much to bear—trying to understand and make sense of the intertwining of my birth mother's death and my first breath. For years after the secret came to light—after it fell on me—I didn't think or talk about my adoption or the details of my childhood. I don't regret it. I was busy. Nikki and I had enough to deal with, raising kids and enjoying grandkids alongside our careers, communities and hobbies.

Then, in 1999, my mother died. Florence. The woman who raised me. The only mother I had ever known before I knew about the other one. Then, suddenly, something shifted in me and in how I spoke with my family members, many of whom had never spoken directly to me about my adoption before. Without Florence there any longer, I found that I was surrounded by family members who were willing to (and even interested in) talking openly about my dark secret.

The seal on my past had been broken, and exploring the complexity of my origin story became a driving force for me. That intensified over time. I started through lighthearted research into a curious character that I learned about in my birth family, and that led to a network of wonderful connections that enriched my life greatly.

The lighthearted part began for me in 2012, when I got very interested in genealogy. When I get interested in something, even if it is lighthearted, I commit. I hired a genealogical researcher—a private investigator pointed

On my 70th Birthday in 2012, visiting the grave of my birth mother.

toward the past, someone willing to dig deep into the archives at the New York Public Library and elsewhere, someone who would pour over microfiche all day long with special magnifying eyeglasses until he found as many details as possible about my past. As he fed me details, I populated Ancestry.com. It is a miraculous site that reflects a miraculous practice—I could sing great praises to the Mormons who created and developed the genealogical database. For a period of time, genealogical research even took over my time on the ham radio. Any question marks in the family tree (When did my great-aunt die? When did my uncle emigrate to the U.S.? Where were the records of my birth mother's hospital stay when I was born and she died?) sent me into a frenzy, inspiring fact-finding missions to graveyards and hospital archives in New York, reaching out to never-met relatives to learn about our common past.

People surfaced as the blurriness receded and the past came into focus. I discovered my Uncle Sam. My father had a brother, one of his many siblings, who obsessed me. He was an artist. I visited with relatives in Vermont. I sought out his paintings. I wrote to the Art Students' League, where he took and taught classes, and I visited their classrooms. Once I had learned as much as possible from my research, I wrote an article for a journal of circus history! Not only had my Uncle Sam painted Franklin Delano Roosevelt, but he had created enormous banners that hung in Madison Square Garden for Ringling Brothers Barnum and Bailey Circus. In fact, these banners hung from the ceiling of Madison Square Garden in March 1942, during a patriotic circus show, an important month for me. It was the month I was born, which meant that it was the month that my mother, his sister-in-law, died.

All this got me thinking about my brother again, enough so that I decided to try to find him. No longer Anesthetist-in-Chief, I had a little more time and space, and into that space questions about my birth family spilled. Was Stanley still alive? Where was he, and what was he doing? As a doctor, I immediately thought about what his medical history might tell me about how I might age, and about the lottery ticket I had passed on to my children. If he was alive, how healthy was he? Should we worry about diabetes? About heart disease? About dementia? Plus there were all the other questions. What had happened to his daughters, my nieces? What had they gone on to do with their lives? Was there anyone in that family (my family!) who was like me?

I found Stanley's number through some Googling, and when I called the number, a woman picked up.

I cleared my throat. "Hello, this is Dr. Zapol, Stanley's um—"

The woman piped up. "Hello?"

I found my words. "Brother. Sorry—this is Stanley's brother. May I speak with him?"

I heard a long silence. "Warren? This is Meryl. Stanley's daughter. I'm sorry to say that Stanley can't speak to you."

My heart sunk. I had missed my opportunity. "I'm sorry too." I said softly. I began to offer my condolences.

She continued. "No," she said. "He's here, I'm visiting with him today, and you can talk to him. He can hear you. But he has had a stroke and he can't really respond."

That information alone scared the shit out of me. Was I sure I wanted to find out more?

I talked to Stanley a bit and then, in the wake of that call, his wife Leah invited me to visit them in Fort Myers, Florida. I impulsively accepted the invitation and promised to visit in the winter, when I would be in Miami for New Year's, catching a sunny break from the Boston winter. But nagging doubts trailed the call. Should I visit? I had a distinct feeling of the tables being turned. Was I the unwelcome stranger/family member, just as I had felt Stanley was to me thirty years before? Was I now the one who had appeared out of nowhere, wanting a relationship and answers?

The doubts passed, or at least most of them did, and on January 3, 2013, I drove to meet my birth relatives. Nikki, David and his daughter Ruthie piled into a rented car with me and we drove a couple of hours from Miami across the Everglades. Ruthie noticed the signs for Barnum and Bailey's circus on the way—as I'd discovered during my research about Uncle Sam, Fort Myers is their winter home—and David bought tickets for a show later that night. I stared out at the alligators as we zoomed by on the highway, hurtling towards a destination I wasn't so sure I wanted to visit. I remember the glint of the sun on the Everglades as we drove west. When we turned off highways into a residential neighborhood, I had another thought. My father Ben had died from a heart attack washing his car on a golf course, and that's where Stanley's house was, on a golf course. It's funny how the pivotal moments of our lives happen during days that are just like all the others, but if we start to unravel the threads we find connections between them.

We took our time getting out of the car. A woman peeked her head out of the front door to look at us. For a moment I wasn't sure if I was in the right place. She then opened the door further, once she had gotten a good look at us. She said, "Oh hi, Warren. Welcome, come in please, Stanley is in the back." I understood that it was Leah, but it was hard to remember in that moment what she had been like before. It had been three decades since we'd been together.

I smiled. Ruthie broke the ice by walking right in the door and introducing herself, and then we all exchanged hugs. I really didn't know what to expect.

We walked to the back through the house, and I saw Stanley as I approached from behind. When he turned, I couldn't tell how much he saw or recognized me, if at all. Perhaps he was angry with me and didn't want to speak. He was sitting in a chair, facing a field behind his house, and had my eyebrows and my olive skin. But he wasn't me. Not at all.

He stood to face me, and my heart skipped a beat. This was my brother. I hadn't seen him in three decades. I stood there and then hugged him.

"Good to see you, Stanley." I said.

He received my hug, but didn't say anything. After some time he mumbled a greeting. The days of exchanging meaningful words had passed, but we were together.

Everyone else crowded around, and my son truly met his uncle for the first time. It was sad that Stanley could no longer communicate. Still, I wasn't there to learn only about him. I was there to learn more about my family.

I turned to Leah. "I've been researching our Uncle Sam," I said, and I pulled out a large printed family tree.

"The painter?" she offered.

"Yes," I said. "Do you know of him?"

"Come back inside," she said, and led us back into a hallway and room with walls filled with paintings. "These are all his."

I looked around, overwhelmed by the abundance of color radiating from the beautiful canvases.

She pointed to a portrait. "That's Millie, your mom."

Portrait of my birth mother, Millie Warshaw, painted by Sam Warshaw.

It was one thing to know that my mom existed, to see her in the old photograph Leah had given me years before, and imagine all the things she did and did not do. What I knew of her life was minimal. She had met my dad, had my brother, died four days after my birth. It was another thing to see a painting by an artist who was family and clearly loved her. That was the closest I could have come to seeing her in life. The brilliant red-orange color of her blouse was unforgettable. The same color was used for her lips and earrings. What was really striking was her look. She was focused, piercing the world with a look off to the side, as if the thing she is looking at just could not be described. Maybe she was looking into the future. Maybe Uncle Sam wanted to believe that she would have an impact long beyond her days. The painting pulled at my heart and tears welled up in my eyes. As I stared, I wanted the painting for myself. I wanted to know my mother. I wanted to look at that painting forever. I stared for a long time before I realized that we were being ushered out of the room.

The next morning we met some of the other family at a Perkins Family Restaurant for pancakes. A teenage girl with long curly red hair and a sparkle in her eyes bounded up to me. "We're relatives! I'm Cait!" We sat together at the booth, crammed in with my family, including Cait's mother Meryl, and one of her brothers, Charles. Caitlin was nineteen at the time, Stanley's grand-daughter, and had been doing very well in school, according to her mother. She was wrapping up high school and looking at a new local college. Her interests lay in science.

Caitlin doted on Ruthie, and at the same time showed curiosity about my career. She was eager to hear stories about my lab, about my work with seals. It's always fun stuff to discuss, and I enjoy sparking enthusiasm in others. I shifted into "lecture mode," as Nikki calls it, grabbing a paper napkin to diagram the blood gas levels in seals as they dive. Usually, that's when kids' eyes glass over. But Caitlin stayed right with it, asking more questions. She was young and hadn't been exposed to this kind of thing before, but she was engaged.

An hour and a half later, as we stood up from lunch to take a photo of this new (or newly assembled) family, I offered Caitlin a job in my lab, if she wanted one. This wasn't the first time I said this to a young, excited person. It wasn't even the first time I had extended an offer within my family. Paul Alfille, my cousin, had worked in the lab and had come with me to Antarctica when he was an undergraduate. David had spent a couple of summers in the lab in Boston as well. Paul and his wonderful wife Laurie work in the same department with me to this day. David and I published seven papers together and continue to collaborate to this day. Still, I meant it. I was overjoyed to meet someone in my birth family who might be like me, who might speak the same language of science and discovery, and who I might be able to nurture, as Florence and Ben Zapol nurtured me.

Caitlin refused at first. She wasn't able to come that summer for some reason. But I kept in touch with her by email and we lined things up so she landed in Boston the following June for an internship at MGH. Nikki met her at the airport and drove her through the early summer sweet air of Cambridge. Back at the house, she installed herself in the basement guest room. I was away at a meeting, but I had left her a stack of papers from the office before I left.

I called home that night. "Read them," I said. "You'll be learning a lot about cardiology. Raj Malhotra will be supervising you this summer. He's smart—an M.D. who trained at Harvard and MIT—and is doing some exciting science in the lab. You'll like him."

This is how we inducted Caitlin into the world of scientific learning and discovery. Away from home for the first time, thousands of miles from her mother, father and brothers, she buckled down and read through the first paper. She called me afterwards. "I'm not sure I'm going to make it through all of these," she said.

"You got this, Cait," I said.

"I don't know…" She trailed off.

"It takes a while to learn the language of a new field, but you'll get there," I said. "Just make a list of questions, mark up the paper and we can talk about them. I'll be back tomorrow night and Raj will be happy to meet you Monday morning. Read before you arrive, it will pay off. You can do it. I know you can."

My general attitude about work in the lab is that it is just like work anywhere else: 90% perspiration, 10% inspiration. Reading science isn't easy, it's work, and the first thing that Cait needed to learn was that it was work she could do. That would help her feel confident learning new concepts and engaging with new science.

By the time I returned on Sunday night, Cait had read the stack of papers. Of course she couldn't understand everything. She was just out of high school. But I really appreciated that she was trying. Could Cait make it in the lab? Could she find her way in the world of healthcare or science? I knew she could, but would she come to believe that too?

Monday morning, we took the T from Cambridge to Boston, walked over to the lab, and went down the bays as I pointed out contraptions and introduced her to the team. Big high-pressure cylinders of gasses—N_2, O_2 and NO—were tied to the end of some benches so that if there was an earthquake they wouldn't fall over and become rockets. Sparking devices made electrical crackling noises as they generated nitric oxide. Computer keys clacked. Tubing hissed and bubbled with a variety of reactions and on every black bench. Clear chambers filled with food, bedding and water for mice awaited their occupants. In each bay a few people typed, talked, or tinkered, looking up to say hello when they could, and working away when they couldn't. The whole place had a buzz to it. It had been my daily environment since I was not much older than Cait, though I knew that it probably felt overwhelming to her.

In the last bay at the end, we found Raj talking to a technician. Raj is a tall and poised figure who dresses formally to see patients, and often runs around with scrubs over his clothes in the lab. He listens before he speaks, speaks concisely and directly, and has a wonderful bedside manner. I could see the anxiety in Cait's eyes as we walked up to him. Once Raj spotted us, he excused himself, and turned to us. "You must be Cait!" he said. "Welcome."

"This is Professor Malhotra," I said. "He is going to be your mentor. He's a great guy. And if you want to learn about hearts, he is the best. You'll be great Cait, just listen—open your mind, and Raj will teach you everything you need to know."

Cait smiled and her eyes relaxed. "Thank you—Dr... Professor... Malhotra, thank you."

"Call me Raj. You'll be working with us this summer and we've got a lot of work to do, so pull up a chair and let's start."

I left Cait with Raj and went back to my office. There's a science to science, knowing how to ask the right questions and look in the right places and do the right work. All of that is transmitted in papers, in discussions, and in lab science in particular, by being together in the same place, trying to do the same work that others are doing around you. That's how you learn each others' skills. That's how you make the science work. It's not about knowing the answer. It's knowing how to find the answer. Raj would teach Cait this and so much more, but first she had to get through the first day.

That night as we headed home, Cait was uncharacteristically quiet. I could see that she was tired. Over dinner I let her relax for a little while before I asked my first question. "How did it go today with Raj?"

"Great," she said. "He's so kind and patient with me." Then she fell silent. "Everyone is so smart and well educated. Harvard, MIT... I'm just a high schooler! I just don't know if I can live up to all these expectations."

"Of course you can," I said, "What did you learn today? I'm sure you have questions. What can I help you with?"

"What is phosphorylation?" she asked.

"Putting phosphoryl groups onto biological molecules. It's important in energy storage, it's the "P" in ATP, and you'll be learning more about it as you read more. I'll share some more papers with you when you get through the first stack."

It was a thrill to see people come in, even someone as young as Cait, without even an undergraduate degree, and start to ask questions. She had only a few short months with us, so I was determined to help her use every single minute of this time valuably. That meant she needed to put in the work, which she did. She worked long hours, staying at the lab past when I'd go home much of the time. She spent weekends reading papers. And it seemed to be paying off. Cait learned how to ask questions that summer—good ones. She turned out to be technically gifted and focused on small animal work, and intellectually she engaged well with Raj and the team.

I like to manage a lab by walking around. I walk down the hallway, stopping in at the benches crowded with stuff and talking with people about what they are doing. When I stopped in to see Cait I peered over her shoulder. It was extraordinary to see my brother's granddaughter sitting there dissecting a mouse's aorta! "Wow," I said. "That's a small vessel!"

Caitlin O'Rourke at a Zapol Lab bench, 2014.

Cait started a little, and then went back to what she was doing. "Yes," she said. "Raj says he can't believe I can get the tube in there, he can't do it. I told him to go take care of some patients. I'll take care of the mice."

A little later in the summer I walked down the hall again to her bench.

"Let's talk." I said. She put down the paper she was reading, and we walked into my office.

"How is it going in the lab, Cait?" I asked.

"Look, Warren," she said. "I don't know if I can hack this. It's really scary, really intimidating. Jessica, who works in the next bay, was just accepted by NASA to be an astronaut. Now I'm working next to an astronaut! I love the lab, but it doesn't get more intimidating than that."

"Cait, you are doing great," I said. "Raj just asked me if you could stay to work here for a year or two. His current technician is leaving, and you have good hands. He could use your support. How about you stay here for the next two years, we hire you to work in the lab, and you can take courses at Harvard Extension School. You can bring up your grades, and you can apply to college from here. I know you can do this."

She didn't blink. Her answer was a single word: "Yes."

I realized talking to Cait that day, that she had grown up in the family I had left when I was a few days old, and I had grown up in a different family. A family where my mother told me that I could do anything: I could fly rockets (which was true, even if they crashed into neighbor's walls); I could make radios (which was true, even if they blew up from time to time); I could lead expeditions (which was true, even if I almost died of malaria). I thought I could do all these things because Florence believed in me. And this was a chance for me and Raj and others in the lab to believe in Cait, and to help her grow.

Cait worked with us for the next year. At a certain point she was an old hand, so Raj decided it was time for her to give a lab meeting. He coached her, though she had probably sat through 40 lab meetings at that point. She knew what it took, and it just took presenting new data clearly so that the team could discuss. She may have been intimidated because the conversations can get heated. It's all for the purpose of science, but sometimes I grill people about their data, finding out what's wrong so we can make it better. I don't mince words—I never have—but I'm not mean either. It was Cait's turn. At the same time that she was working on her presentation, she had college applications due, so I was a bit nervous for her. I noticed that her skin was beginning to break out. Would she choke under pressure? I didn't think so. Wednesday Lab meeting in the Zapol Lab was held in a big conference room with about 20 chairs around a series of tables. There were big posters on the wall, mostly of Antarctic scenes—photographs of seals in ice holes, prints of historical ships crushed in ice. The letter President Bush sent appointing me to the U.S. Arctic Research Commission. I liked these images—they got people out of the day-to-day of Boston and started them thinking creatively.

The Zapol Lab was actually the Zapol-Berra-Bloch-Buys-Ichinose-Malhotra Lab, or something like that. We had a bunch of professors in the group, and they all ran their own labs, but we all got together to share ideas. The main lab meetings were big. They were places to hash out ideas. We went around the room, got updates from everyone, and listened to a few people present their data. Sometimes it lasted an hour, other times it went on for several hours.

Cait stood up, took a deep breath, turned bright red to match her reddish hair—her father was Irish—and launched into her talk about her project. We didn't need to grill her, but we did ask some hard questions, and she did great. I was so proud of her, and more importantly, she saw she could do it.

She stayed on for three years in the lab, published a few great papers with Raj, and ultimately graduated from Boston University with a degree in biology. She came back to the lab for a postgraduate year, and today she is working in a pharmacy back in Florida. Raj and I hear from her from time to time. She is now part of my family, as well as the lab community. I hope that her time with us in Boston helped her find her way.

It makes me think of Uncle Sam, Cait's great-great-uncle, and how he found his passion in painting. He did it beautifully, on a national scale, and was so very different from his brother, my father, who was dishonorably discharged from military duty in the South Pacific. We all find our way with the family upbringing that we have. Cait walked across family lines and joined me in my sensibility: shaped by my two mothers, my many families, somehow driven to practice science and uncover the mysteries of the universe. I am so proud to have found a way to bring the members of my families—my biological parents, my adoptive parents, my own children, the extended families and the intellectual colleagues who are as close to me as blood relatives—all together in the scientific pursuits I have chased now for over half a century.

CHAPTER 13

The Patient

∗ ∗ ∗

IN LATE SUMMER 2015, I was in Anchorage and Nome, Alaska for a meeting of the Arctic Research Commission. Traveling from Boston to Alaska is a strenuous trip in the best of times. I usually try to stay there long enough to recover before returning, even taking an extra day to go fishing with my friends on the Commission. This time, though, when I made it back to Boston, I was tired, and worse than usual: I was short of breath, coughing, and my chest hurt. The differential diagnosis I made on myself scared the hell out of me. I told Nikki that I either had pneumonia or a pulmonary embolism. I did not mention cancer because I knew it would scare the hell out of her, too. When I described my symptoms to my primary care doctor, he ordered me to come to MGH Monday morning. I did. He told me then that it was most likely non-small cell lung cancer, NSCLC. It is the most common type of lung cancer, one that is often found in non-smokers, and until now, one without any known cure.

The good news about being at MGH for so long was that I kind of knew where the great doctors were. Whenever I needed a doctor, I could usually find one and get treated and get an answer that I trusted, an answer that was probably right. On the other hand, a process like this is humbling. It makes you realize both how good the process is and how bad it is, how good we are as doctors and scientists and how bad we are. How good we are because we use modern medicine and the latest drugs to conquer old killers, or at least set them back on their heels. And how bad we are because we're not exactly sure that our treatments are a good fit for our patients. Every patient is an "n equals one," meaning that we have never seen their situation before, not exactly, and that requires us both to know what we are doing and to rethink what we know and what we do next.

That day when I went into MGH and came out with a cancer diagnosis put me on the other side of that equation. I started experiencing that reality as a patient, which was a completely different existence from the (somewhat) dispassionate distance I could maintain as a physician researcher. Uncertainty about the course of this illness, suddenly, was not an academic challenge but rather an existential reality. It was a game changer for me and my family. Nikki's father had died of kidney cancer two days after she and I were married. That memory cast dark shadows over the prospect of illness throughout our lives together, and then that shadow settled in over me.

Cancer. Lung cancer. I grew up as a doctor in an era when everybody died from this disease. I saw many of those patients, took care of them, or anesthetized them, and watched them struggle for breath as they declined. It was awful. But there was nothing I could do for them beyond that. Back then, we used crude drugs, few of which worked. There was no crizotinib, no erlotinib, which are tyrosine kinase inhibitors, or drugs that inhibit phosphorylation of proteins. Remember the question Cait had for me? In this case, they disrupt the instructions that lead to cancer cells multiplying. That was what was high on my mind when I first got my diagnosis in early September, 2015, and why I declared at the time that I would not be around for Thanksgiving. I knew it for a fact and I was afraid, afraid of an increasingly painful end that would not go quickly enough.

My doctors had a different view. They approached me as if I were harboring a challenging genetic puzzle. Indeed, my cancer's genes revealed a familiar puzzle piece, an *EGFR* mutation, or a change in the epidermal growth factor receptor gene which causes the cancer to grow. That was good news, they told me, because it was the most common NSCLC mutation and therefore the one for which there were the greatest number of new drug treatments. In addition the analysis revealed a *MET* amplification, which also makes the cancer grow quickly. The truly unusual aspect of the puzzle was that both *EGFR* and *MET* mutations were present from the start. Ordinarily, to a physician-scientist, "unusual" presents an opportunity.

That was six years ago, and since then for the first time in my life I was featured in a scientific journal as a patient, not a doctor or a scientist. My double mutation cancer, and the brilliant science behind how my doctors solved that puzzle and kept the cancer at bay made for good scientific news. For a while, lung scans showed the treatments were working. Then they showed the cancer was advancing. In 2016, tumors popped up in my brain. Miraculously, at each of those times, my doctors ferreted out the very latest drugs or radiation treatment that beat it back. I came to believe that they always had a rabbit to pull out of the hat.

Though I have had six great years living with cancer, the process of picking a path through the last six years of therapy was stressful. I had a cancer that responded to an evolving combination of new treatments never before tested. As much as I wanted to be a patient in the hands of expert physicians, they could not tell me what to do. I had to choose from a range of bad options: accepting known drugs with known side effects that had little chance of beating back the cancer or trying experimental combinations which might, just might, do better… or worse. Asking my physicians for their opinions felt like the partially blind leading the partially blind. At Nikki's insistence, I reached out for another opinion: two of them, in fact. Neither expert I spoke to said: "I want you to do A, B, or C." Rather, the message I got was, "Well, you could do A, B, or C." The second expert felt a newer but less-well-understood drug held the most promise. It was up to me to choose the least bad option. I chose the new drug and I got lucky.

Physical therapy was also key. "You can do the bike—you can do it," my MGH physical therapist said. "Your heart rate's ok, your pressure's ok, go do it." She was terrific. I joined the MGH gym, and went to Florida during the winter with Nikki, biking, and biking more and biking more and biking more.

While I was there, Liza had an idea. "Dad," she said, "let's do The Pan-Mass Challenge." The PMC, a charity bike-a-thon that has, since 1980, raised money for cancer research, is the largest such event in the country. In 2016, 6,260 riders raised $47 million.

"I can't do it," I said. "I can't possibly do it."

David chimed in "Dad, let's do it."

I relented. For the first few months of training, I did not ask for any donations (PMC requires a sizable amount to participate) because I was afraid that I might not live long enough to do it.

In August 2016, David, Liza and I (I wore my "Survivor" biking costume) rode the 90 miles from Bourne to Provincetown. As we started out in the dark,

I began shivering uncontrollably, shaking my handlebars. My body couldn't generate enough heat, not until David warmed me with his jacket. Then, the sun decided to give us an orange glowing sunrise show about 5:45 am as we reached the Sandwich oil tanks. We had to stop and get a photo of the three of us—that is the most memorable sunrise of my life.

Miraculously, the rest of the ride went smoothly. I happily rode alongside David, or Liza, chatting away. We rode through the sweet town of Dennis, and along the wonderful bike paths of Wellfleet where not so long ago, I taught Liza to ride a bike. We rode over the cliffs, and onto the sandy highway in Truro where I narrowly avoided a fall. And finally, we climbed the dunes of Provincetown, and decided to go the extra long route to the finish line.

Finally, we all made it intact, greeted by high-fives, hugs and a wonderful handmade sign by Nikki, Ruthie, Elliot, Juno and DT. I earned my Sam Adams in the celebration tent, where they blasted Bruce Springsteen and I rested my weary feet on a chair, toasting an incredible ride.

The Pan Mass Challenge was a turning point for me. I realized that if I could bike 90 miles, I could probably re-integrate myself into the world. I could socialize again, work again, travel again. That is how I feel now. Nothing is hurting me. I know my life span is uncertain—but isn't everyone's? In many ways the drugs have converted my thinking about cancer as an acute disease to a more chronic disease. These days it could be more like having a leaky heart valve or prostate cancer. Yes, it may do you in eventually. Those billions of cells are nasty little brains trying to figure out how to do you in. But there are always exceptions that buy more time, sometimes in large denominations. Get on with it!

That has been the drum beat that has kept me going. I have returned my focus to moving the frontiers of science forward with the goal of bringing new treatments to patients everywhere. I have traveled in this country, as well as to Europe and Asia several times to press on with the commercialization of inhaled nitric oxide and to make it available and affordable. I've welcomed more fellows to my lab from Germany, Italy, China and Japan. I've had the unique, often wonderful, sometimes challenging, experience of working with my son, David.

Have I noticed any changes? Only that as a patient I have become more impatient. In 2020, the COVID-19 pandemic spread across the planet. A health problem of that magnitude took my mind off my own health issues, to some degree. I am finding that I cannot tolerate the politics and bureaucracy that are interfering with the use of nitric oxide as an anti-virucidal treatment. That's more and more the case about more and more things. I want to know *now* that nitric oxide will become available and affordable in the developing world. I want to know *now* that the research in my lab to find a treatment for malaria will continue to be supported. I want to know *now* that the outstanding researchers whose careers I have promoted are being given their due in our competitive academic environment, so that they can continue to thrive and make their mark.

And, of course, my thoughts turn to my family. I fear they have never gotten as much attention from me as they deserve, and even now I confess I cannot promise to change my ways. Nikki has tolerated me and my personality—the passions and the blind spots both—for more than 50 years, with love. I would not have been able to accomplish a fraction of what I have without her support, and that includes raising two terrific children. Through it all, I feel I wasn't there enough for them or for her. Still, I have enjoyed including family members in my peripatetic life. Liza was born when I was on sabbatical in England. David worked as my lab tech in Antarctica and met his future wife there. Liza served as my companion on a Harvard-sponsored Antarctic expedition. Ruthie, my granddaughter, went to China and Japan with me. Nikki accompanied me on several Antarctic expeditions where she lectured on Antarctica's legal status. I dream one day of bringing Elliot and Juno with me, too. But my luck may run out. I hope they forgive me for not doing more.

Justin Gainor, my brilliant oncologist at MGH, told me six years ago that there was no known cure for my cancer. The rollercoaster of these past years has at times made me wonder whether I might prove him wrong. But to hedge my bets, perhaps you will excuse me if I go into a time machine. I'm setting the dials for ten years in the future. Certain areas are going forward at a good speed, like the area of my own disease treatment. But it's not there yet. They're learning, and they will continue to learn; they'll be very good at it in a decade. They'll say, "Gee I wish I had done that," as we all do.

* * *

Warren died on December 14, 2021, five days after walking on his own from the drop-off at the Yawkey Center for Outpatient Care at MGH into his doctor's appointment. Dr. Gainor immediately determined that Warren was doing poorly. His oxygen saturation had plummeted, and his lungs were filled with fluid, again. How was he walking and breathing, unassisted by supplementary oxygen? Dr. Gainor arranged for him to be transported to the emergency room. Hours later he was transferred to a private room overlooking the Charles River under the care of his intensive care colleagues. He agreed to breathe his own medicine, nitric oxide, when all else was failing. He noted that NO improved his oxygenation, but not enough to turn the tide. He smiled, though, declaring, "As billed, it is odorless and colorless." He and his colleagues had used that exact language to argue with the FDA, that NO could be effectively blinded in designing clinical trials.

Warren declined rapidly, but continued teaching and challenging everyone through his last hours. Here are some excerpts from his recorded final conversations with his doctors and friends: his final grand rounds.

December 13, 2021, Hospital room, MGH Phillips House
Warren:

> Possible treatments keep popping up. My doctor is quite certain this latest one would just be Scotch tape. But… but would it work here? It's too early to know. Some of us are smart, and we know where the field is going—and some of us ain't. As a senior scientist, totally untutored in modern molecular biology, and what possibilities are, I determine a certain amount of, uh, discouragement in my doctor. He had a good run with me for five or six years.

> But…you have to believe. And in fields where change occurs, it will be slow.

> You might dream of novel therapies, but they're dreams until proven. My favorite words are from the FDA: Have patience, Warren. You'll get there eventually. Took me nine years 'till we did the first babies

here with pulmonary hypertension.... It took nine years to get all the approvals. And you know, we did our first ECMOs here. One at MGH, one at the Brigham, in 1970. And then I used to do one a month, and try and figure out what I did wrong, patients bled... And it took until 2010 before we were doing a hundred a year at MGH.

Warren's palliative care consult asks him what he thinks is going on with him now.

Warren:

Oh, I think several processes. I can't prove it. But, you know, the process of inflammatory response is such a complex field. My left lung, which is where it supposedly originated, probably does have an inflammatory component. Is it, as an oncologist would say, post-radiation pneumonitis? How much of it is old? How much of it is new? You know, obviously, the destruction of the anatomical creature, the lung, gets very complex when you use various therapies. Our radiation oncologist suggested we now chase the "little dots." There's some chance with that, too. We just don't know. Could it be pneumonia? Well, there's my colleague cardiologist. Much like me, an intensivist, much more believing in the interaction of infection with disease. And just because he's a cardiologist doesn't mean he can't be infected. Just because you're an oncologist, you could be pretty well infected. Think of the world. Its main problem is infection. So, it's complicated. I tend to work both with other physicians, and other specialists. I never, never believed that my specialty was *the* specialty. You stand on the shoulders of giants, and so keeping one's mind open is very important,

I have no idea whether these cells are gonna eat me or not, or take my life. I don't think anyone will know till they do the final autopsy what the hell was going on, and even then that may be hard to pin down. Because, as you know, they can give you a list of things that are wrong with you, but the interplay between each one of those things—is so complex.

I'm talking too much… I'm sorry.

The palliative care doctor asks Warren if he is ready to begin withdrawal of treatment.

Warren:

> I would keep carrying on. I mean, I put all the pluses in one box, all the minuses in one box. I think about my family, what they have to put up with, me, and my hopes.

> Why am I in this world? To teach people. To let them understand. And, for many of the foreigners who come here in droves: helping them to see the confusion in advancing medicine. It is very difficult. But fun. And humbling, Just think about it. The field turns over so quickly.

> So, why am I here? I'm here to make things understandable. And yes, I don't know oncology, but I know probably how the lung works, or tries to work. So my own plan is to collect the facts. If the facts go down, go south, that won't be nice.

> I've been trying to doubt what anyone thinks.

December 14, 2021

Dr. Gainor enters Warren's room around 12:30 pm. He sits at Warren's bedside and explains the data behind his conclusion that the cancer recently returned with a vengeance, now presenting insurmountable obstacles.

> Gainor ends his analysis, saying softly: " I really wish it weren't the case."

> Warren, barely audible, turns to Nikki: "They win sometimes…"

Nikki chokes back tears: "We won with you for six years, dear."

Warren: "Yeah."

Warren raises his right arm and makes the gesture of a scissors. Barely audible, he seems to say "*genug*" which means "enough" in Yiddish.

The room empties except for Warren, Nikki, David and Liza.

At 9:37 pm, Warren is pronounced dead.

Account of Warren Zapol's final days by Nikki Zapol
September 2024

At Warren's memorial service in the Ether Dome on December 16, 2021, his family gathered in front of the Prosperi painting in which Warren posed as Morton, holding the ether inhaler. From left: Diana Laird, Ruthie Zapol, David Zapol, Elliot Zapol, Juno Townsend, Nikki Zapol, David Townsend and Liza Zapol.

Acknowledgements

*　*　*

There are vast gaps in this memoir, many more adventures to be told, and much, much more to say. I am racing against the clock now. I must therefore count on my dear ones to take what I have already written and bring it across the finish line to publication.

Nikki, my devoted wife of now fifty-three years, my son, David, and my daughter, Liza, are steadfast in their belief in me and my work. Their support and love have been critical during the past challenging years of my illness. I owe them the deepest gratitude. As I approach eighty, they have helped me sustain my momentum, clarify my memory, and put it all into words.

This memoir would not have taken form, nor would my career have been as productive without the constant and steadfast help of my assistant Missy Flynn.

My friends and acquaintances have offered advice on this manuscript and their contributions have helped to shape the document you read.

So many people have played important roles in these stories, but not all have been named here. Forgive me for this, for you were instrumental. In addition, those of you with whom I was lucky enough to share these events may remember them differently. I encourage you to tell your own version. Please keep telling these stories.

Cambridge, Massachusetts, 2016.
Photograph by Jon Lowenstein.

APPENDIX
Fellows and Technicians

* * *

Fellows are the basis of my work, my teaching and mentorship efforts. Many taught me about their countries of origin, and many are the clear basis of my scientific advancements. Therefore I will list fellows and technicians whose salary I paid by their countries of origin and first names. I apologize to anyone whose name I have overlooked.

USA
Adam Sapirstein
David Polaner
Erni A. Kreil
Hemanth Baboolal
Jesse D. Roberts
Jessica Meir
Michael T. Snider
Orlando C. Kirton
Richard I. Whyte
Robert Schneider
Roger Snow
Ryan Carroll
Virginia Schmidt
William E. Hurford

Germany
Alexandra Holzmann
Andrea Steinbicker
Anna Fischbach
Bodil Petersen
Claire Manktelow
Jan A. Graw
Joerg Weimann
Martin Wepler
Matthias Derwall
Matthew Kurrek
Peter Radermacher
Roland Francis
Steffen B. Wiegand
Stefan Muenster
Wolfgang Steudel

Italy
Ester Spagnoli
Fabrizio Michelassi
Lorenzo Berra
Luca Bigatello
Luca Zazzeron
Luciano Landa
Pietro Caironi
Riccardo Pinciroli

Japan
Akito Shimouchi
Akito Nakagawa
Fumito Ichinose
Hirosuke Kobayashi
Koichi Kobayashi
Noriko Kawai
Ryuji Hataishi
Yasuko Nagasaka

China
Binglan Yu
Chong Lei (Crystal)
Tong-Yan Chen
Yandong Jiang

France
Christophe Adrie
Gilles Montalescot
Marie-Dominique
   Fratacci
Patrick Dupuy

UK
Beeshma Rajagopalan
George Collee
Robert Grange
Roger D. Hill

Canada
Allyson Hindle
Robert Boucher
Francine Lui

Austria
David Baron
Roman Ullrich
Marianne Winkler

Israel
Eran Geller
Yehuda Raveh

Russia
Arkadi Beloiartsev
Oleg Evgenov

Spain
Antonio Torres
Irene Rovira

Chile
Gian Paolo Volpato

Denmark
Peter Huttemeier

Greece
Katerina Vaporidi

India
Mohd Shahid

Mexico
Guillermo Castorena

Poland
Marek Skoskiewicz

Sweden
Claes Frostell

Switzerland
Denis Morel

Technicians
Caitlin O'Rourke
David Zapol
Kevin Stanek
Elizabeth Greene
Melahat Samali
Paul Alfille
Scott Tolle
Thomas R. Wonders

About the Author

$*$ $*$ $*$

Warren M. Zapol, MD (1942–2021) was Anesthetist-in-Chief at Massachusetts General Hospital, a founder of the MGH Anesthesia Center for Critical Care Research, Reginald Jenney Professor of Anesthesia at Harvard Medical School, and a faculty member of the Harvard-MIT Program in Health Sciences and Technology. An exceptional scientist, mentor, doctor and engineer, Dr. Zapol was an inventor of the therapeutic use of inhaled nitric oxide, which has been administered to hundreds of thousands of patients, and is a live-saving world-wide standard of care for "blue babies." Dr. Zapol's research and medical missions spanned the globe. He led nine Antarctic expeditions to study how Weddell seals hold their breath and avoid the bends. In 2008, he was appointed by President George W. Bush and reappointed in 2012 by President Barack Obama to the U.S. Arctic Research Commission. He also served on the Polar Research Board of the National Academies of Sciences. He was a member of the National Academy of Medicine and in 2003 was awarded the Inventor of the Year Award of the Intellectual Property Owners Association. The Zapol Glacier, named after him, flows from the largest mountain in Antarctica.

A full oral history with Warren Zapol conducted by Liza Zapol can be found in the Harvard Countway Library's Center for the History of Medicine.

Find out more at: DrAdventure.org

www.ingramcontent.com/pod-product-compliance
Lightning Source LLC
Chambersburg PA
CBHW051615120626
46551CB00014B/1802